Accenting Lipid Peroxidation

Edited by Pınar Atukeren

Published in London, United Kingdom

IntechOpen

Supporting open minds since 2005

Accenting Lipid Peroxidation
http://dx.doi.org/10.5772/intechopen.92399
Edited by Pınar Atukeren

Contributors
Harnavi Harun, Arnold N. Onyango, Kale Ezekiel Oluwafemi Ezekiel, Raju Ratan Ratan Wadekar, Kalpana Sabanna Patil, Alfred Mavondo-Nyajena Mukuwa Alfred Mukuwa Greanious, Francis Farai Chikuse, Paul Makoni, Alfred Sibanda, Nesisa Ncube, Delton Dube, Suzan Onur Yaman, Adnan Ayhancı, Abdulsamed Kükürt, Volkan Gelen, Hacı Ahmet Deveci, Mahmut Karapehlivan, Ömer Faruk Başer

First published in London, United Kingdom, 2021 by IntechOpen
IntechOpen is the global imprint of INTECHOPEN LIMITED, registered in England and Wales, registration number: 11086078, 5 Princes Gate Court, London, SW7 2QJ, United Kingdom
Printed in Croatia

British Library Cataloguing-in-Publication Data
A catalogue record for this book is available from the British Library

Additional hard and PDF copies can be obtained from orders@intechopen.com

Accenting Lipid Peroxidation
Edited by Pınar Atukeren
p. cm.
Print ISBN 978-1-83968-825-6
Online ISBN 978-1-83968-826-3
eBook (PDF) ISBN 978-1-83968-827-0

For EU product safety concerns:
IN TECH d.o.o., Prolaz Marije Krucifikse Kozulić 3, 51000 Rijeka, Croatia, info@intechopen.com or visit our website at intechopen.com.

Meet the editor

Dr. Pınar Atukeren is the Dean of the Faculty of Pharmacy of Istanbul University-Cerrahpaşa, Istanbul, Turkey. She is also teaching in the Department of Medical Biochemistry of Cerrahpaşa Medical Faculty, in the same university, as a senior professor from where she obtained her Ph.D.degree in 2005. She had been awarded the National Research Grant for Successful Students from the Scientific and Technical Research Council of Turkey. Her research interests are oxidative and nitrosative stress, free radicals and antioxidants, natural therapeutics, neurodegenerative diseases, brain tumors, and aging. Dr. Atukeren has published many papers in indexed journals and edited and authored both international and national books and book chapters. She is a reviewer for many international peer-reviewed journals. She has attended many conferences and workshops and is involved in many training courses and consultations. She is also the head of the Animal Welfare and Ethics Committee and a member of the Scientific Research Projects Coordination Unit of Istanbul University-Cerrahpaşa. At the community level, Dr. Atukeren is an active member in several governmental and non-governmental societies and committees.

Contents

Preface

Oxidative stress is the imbalance between the production and the accumulation of free radical molecules, which are reactive oxygen species, in cells causing certain physiological effects and various diseases. Lipid peroxidation is a free radical-mediated chain of reactions involving the formation of lipid radicals and rearrangement of double bonds in unsaturated lipids. Lipid peroxidation is initiated and enhanced via several toxic products resulting in oxidative damage to polyunsaturated fatty acids in the cell membranes, which are the most common targets of cellular components.

When biological membranes are destroyed, a variety of breakdown products are produced that lead to cell death. Lipid peroxidation end products are cytotoxic products that cause covalent modification in cellular macromolecules. These macromolecules play important roles both in the development and the progression of many diseases via altering redox homeostasis.

This book provides information about lipid peroxidation and will help guide the development of potential therapeutic approaches to control the production of free radical-mediated generation of lipid peroxidation to combat human disease.

Pınar Atukeren
Cerrahpasa Medical Faculty,
Department of Medical Biochemistry,
Istanbul University Cerrahpaşa,
Istanbul, Turkey

Chapter 1

Lipid Peroxidation

Suzan Onur Yaman and Adnan Ayhanci

Abstract

Lipid peroxidation (LPO) is initiated by the attack of free radicals (eg OH ·, O2- and H2O2) on cellular or organelle membranes phospholipids or polyunsaturated fatty acids (PUFA), and with the formation of various types of aldehydes, ketones, alkanes, carboxylic acids and polymerization products. It is an autoxidation process that results. These products are highly reactive with other cellular components and serve as biological markers of LPO. Malondialdehyde (MDA), a toxic aldehyde end product of LPO, causes structural changes that mediate its oxidation, such as fragmentation, modification, and aggregation, especially in DNA and protein. The excessive binding of these reactive aldehydes to cellular proteins alters membrane permeability and electrolyte balance. Degradation of proteins leads to progressive degradation of the biological system mediated by oxidative stress. The chain reaction (CR) of LPO is initiated by the attack of free radicals on the PUFA of the cell membrane to form a carbon centered radical (R*). The O2 · - radical attacks the other lipid molecule to form lipid hydroperoxide (ROOH), thereby spreading the CR and forming the lipid peroxyl radical (ROO). These lipid hydroperoxides severely inhibit membrane functionality by allowing ions such as increased hardness and calcium to leak through the membrane. Damage to the lipid membrane and macromolecule oxidation can result in activation of necrotic or apoptotic tissue death pathways if severe enough.

Keywords: lipid peroxidation, free radical, malondialdehyde

1. Introduction

Free radicals are formed during the reactions required for the maintenance of normal metabolism and energy formation in biological systems. Under normal conditions, the most significant source of free radicals in cells is the leakage of electrons into molecular oxygen from electron flow in the mitochondria and endoplasmic reticulum during oxidative respiration. The superoxide anion formed in this way is converted into hydrogen peroxide, a reactive oxygen type. Hydrogen peroxide forms the peroxyl radical, which is the most reactive radical type in the organism in the presence of transition metal ions. When free radicals cannot be removed from the environment, they cause damage at the cell, tissue, and organ level by disrupting the structure of biomolecules such as lipids, proteins, and nucleic acids due to their high reactivity. Lipid peroxidation begins when polyunsaturated fatty acids (PUFAs), which are in the structure of membrane lipids in cells, are affected by free radicals. If peroxyl radicals are not cleaned, a chain reaction starts affecting the intact PUFA. Lipid peroxidation is damaging as it is a self-sustaining chain reaction. When lipid peroxides (LOOH) are broken down, aldehydes are formed, many of which are biologically active. These compounds are either metabolized at the cellular level or diffuse from their initial domains and spread damage to other parts of the cell. When peroxidation of fatty acids containing three or more double bonds. Malondialdehyde (MDA) arises following the peroxidation of fatty acids comprised of three or more double bonds and appears in blood and urine. Due to its ability to corelate well with the degree of which lipid peroxidation occurs despite not being a specific or quantitative indicator of fatty acid oxidation, the measurement of MDA in biological material is used as an indicator of lipid peroxide levels. Nonenzymatic lipid peroxidation is a very harmful chain reaction. It both damages the membrane structure directly and also damages other cell components indirectly with the reactive aldehydes it produces. Thus, it causes tissue damage and many subsequent diseases.

2. Free radicals

According to quantum chemistry, only two electrons can enter the structure of a bond together. Electron pairs exist in a very stable state. Electrons in the human body exist almost entirely in electron pairs. When a bond breaks, the two electrons are either separated but remain in the same atom or both remain in the atom separately. If they remain together, the atom formed becomes an ion, and when they leave, the atom formed becomes a free radical [1]. Atoms or compounds that contain the unpaired electron in their final orbital are defined as free radicals. In other words, they are atoms or molecules that have an open electron shell configuration and contain an odd number of electrons in their structure [2]. The term Reactive Oxygen Species (ROS) is more commonly used in place of the term free oxygen radical as it includes molecules that are radical and are not actually radical, but that cause the formation of oxygen radicals with their reactions [1].

Despite oxygen being crucial for life, in some cases it can also damage cells. This damage is caused by increased oxygen-induced ROS production. The amounts of ROS produced under normal physiological conditions do not exceed the capacity of the natural antioxidant defense systems in the body. ROSs are chemical derivatives with unpaired high energy electrons in their outer orbits. In order to stabilize, ROS interact with any molecule they can find in their vicinity and exchange electrons. Molecules that react with free radicals turn into free radicals and initiate the damage chain reaction. These radicals react with organic and inorganic chemicals such as protein, lipid, and carbohydrate. When radicals occur in cells, they react with nucleic acids and various membrane molecules and break them down. While radicals affect intracellular organelles, they also create distant effects by passing to the extracellular compartment [1]. Although oxygen is crucial for human life, some ROS that occur during normal metabolism have the potential to cause great harm to the body. Compared to normal oxygen molecules, ROS, which are mostly composed of free radicals, appear as oxygen forms with higher chemical reactivity [3].

Free radicals are any atom or molecule with one or more unpaired electrons produced in many physiological or pathological conditions. These molecules, also known as oxidant molecules or reactive oxygen particles, easily exchange electrons with other molecules [4]. A compound can return to a free radical by losing an electron (reduction) or gaining an additional electron (oxidation). Free radicals can be part of a larger structure, as well as in immobile or small and freely spreading species [5–8]. Free radicals are frequently produced by the mitochondria during the body's normal use of oxygen. These free radicals, which are formed as a result of energy production, can change the structure of lipids, proteins, and nucleic acids. Free radicals are produced from many endogenous and exogenous sources as well as mitochondria and cause a variety of damage alongside their benefits. The benefits of free radicals only occur when they are of low concentration. Low concentration free radicals are involved in the activation of cellular signals such as calcium release from intracellular stores, and the activation of tyrosine phosphating and growth factor signals, along with defense functions such as defense against infections, the killing cancer cells and detoxification of xenobiotics [9].

3. Ways in which free radicals form in the cell

Common biochemical events, such as those which occur during normal respiration, cause reduction–oxidation (redox) reactions. The molecular oxygen in mitochondria is gradually diminished with the adding of four electrons to form water. Several toxic intermediate derivatives occur during this event. These include superoxide radicals (O^{-2}), hydrogen peroxide (H_2O_2) and hydroxyl (OH^-). In addition, some intracellular oxidases such as xanthine oxidase directly form superoxide radicals as a result of their activities. It catalyzes the formation of free radicals (Fe^{+++} $H_2O_2 \rightarrow Fe^{++++}$ $OH^- + OH^-$) as in the Fenton reaction by exchanging free electrons during some intracellular reactions in exchange metals such as copper and iron. To play a part in the Fenton reaction, the intracellular free iron, occuring in the ferric state (Fe^{+++}), must initially be reduced to its ferrous (Fe^{++}) form. Iron and superoxide are both required for maximum oxidative cell damage as the reduction is amplified by the superoxide ion.

By absorption of radiant energy (such as ultraviolet light, X-rays): For example, water can be hydrolyzed to hydroxyl (OH^-) and hydrogen (H^+) free radicals with ionizing radiation.

By the intracellular enzymatic metabolism of external chemicals or drugs: For example, CCl_3 free radical is formed as a result of the intracellular metabolism of carbon tetrachloride (CCl_4).

Nitric oxide (NO), an important chemical mediator normally synthesized in various cell types, reacts with oxygen, especially non-radical peroxynitrite, a type of free oxygen that inhibits mitochondrial respiration, as well as the radical nitrogen dioxide (NO2) and nitrogen trioxide (NO3) [1]. Common free radicals, their symbols and identities are shown in the table below (**Table 1**):

Free radical	Symbol	Identity
Hydrogen	H	The simplest radical.
Superoxide	$O_2^{\cdot -}$	The first intermediate product of oxygen metabolism.
Hydroxyl	$OH^{\cdot}$	The most toxic (reactive) oxygen metabolite radical.
Hydrogen peroxide	H_2O_2	Reactivity is very low, molecular damage ability is poor.
Singlet oxygen	O_2^-	Strong oxidative form of oxygen with fast half-life.
Perhydroxy radical	$HO_2^{\cdot}$	Rapidly dissolves in lipids and increases lipid peroxidation.
Peroxide radical	ROO^-	is less effective than perhydroxyl, localized to lipids.
Trichloromethyl	CCl_3	CCl_4 is a radical produced in the liver, the product of metabolism.
Thiyl radical	$RS^{\cdot}$	General name for sulfurous and unpaired electron-containing species.
Alkoxyl	$RO^{\cdot}$	Oxygen metabolite produced by the breakdown of organic peroxides.
Nitrogen oxide	NO	is produced in vivo from the amino acid NO L- arginine.
Nitrogen dioxide	NO_2	is produced by the reaction of NO with oxygen [10].

Table 1.
Free radicals, symbols, and identities.

The most important free radicals that occur are:

- Superoxide radical (O_2^-).
- Hydrogen peroxide (H_2O_2).
- Hydroxyl radical (HO^-).
- Singlet oxygen ($O_2\uparrow\downarrow$) [11, 12].

3.1 Superoxide radicals (O_2^-)

The superoxide anion radical (O_2^-) is produced by the single electron reduction of oxygen which acts as an intermediate in a number of biochemical reactions in body [13] and is a weak oxidant that cannot cause serious cell damage by itself.

However, it may lead to the initiation of a series of reactions that can lead to oxidative stress [6, 14, 15]. One of the main points of superoxide production is Coenzyme Q, and this anion is formed at other points in the electron transport chain as well as in the mitochondrial electron transport chain. Another ROS is produced by the O_2^- radical, which does not leak far from where it originates [12, 16].

$$\begin{array}{l} O_2 + é \rightarrow O_2^- \text{ (superoxide radical)} \\ H_2O_2 + O_2 \rightarrow HO^- + OH^- + O_2 \end{array} \quad (1)$$

The OH^- radicals produced are highly reactive and can cause significant damage by reacting with structures such as DNA [6, 17, 18].

The half-lives of superoxide radicals that produce H_2O_2 and oxygen by the dismutation reaction are quite short. This reaction occurs spontaneously and is catalyzed by the Superoxide Dismutase (SOD) enzyme [6].

$$O_2^- + O_2^- + 2H + SOD \rightarrow H_2O_2 + O_2 \quad (2)$$

In natural conditions, O_2^- can be produced in muscle tissues in a variety of ways. One of the sources of O_2 in muscle tissues are various components of the electron transport chain in mitochondria, such as NADPH-linked dehydrogenase and ubiquinone, which can leak electrons into O_2. Autoxidation of heme proteins [19, 20] and metabolic enzymes such as xanthine oxidase [21] are other sources of O_2. With the ingestion of bacteria, the activation of several leukocytes in the vasculature of the muscle tissue causes the production of O_2^-, one of the major bactericides [22].

3.2 Hydrogen peroxide (H_2O_2)

Aerobic cells naturally contain low concentrations of hydrogen peroxide (H_2O_2) as a metabolite. In an O_2 forming system, it is expected to give H_2O_2 catalyzed by nonenzymatic or superoxide dismutase (SOD) [23]. Although it is not free radical, hydrogen peroxide reacts with a transition metal (e.g. Fe^{+2}) to form a free radical [16].

H_2O_2 production has been detected in mitochondria, microsomes, peroxisomes and phagocytic cells. Also, many enzymes, such as xanthine oxidase, aldehyde oxidase, urate oxidase, glucose oxidase, glycolate oxidase, and D-amino acid oxidase, can directly produce H_2O_2 [23]. It has been reported that H_2O_2 is produced at a hemoglobin rate of approximately 3.9 × 10-9 M/hg and the concentration of H_2O_2 in red blood cells in a steady state is 2 × 10-10 M [24]. It has been reported that H_2O_2 formed during oxidation of oxymyoglobin plays an important role in lipid peroxidation [25]. Furthermore, it was reported that turkey muscle tissues stored at 37°C for 30 minutes produced approximately 14.0 nmol H_2O_2 per gram fresh weight, and its formation increased with storage at 4°C [26].

H_2O_2, which lacks unpaired electrons, is not a radical and, unlike charged O2, shows limited reactivity and permeability to the membrane [27]. Nevertheless, H_2O_2 can have devastating effects by generating more reactive species such as OH by catalysis of Fe (II) [28]. In addition, H_2O_2, depending on its concentration, can denature heme proteins to release iron and heme group or to convert heme protein to ferryl or perferryl radical [20].

3.3 Hydroxyl radicals (HO^-)

The hydroxyl radical (OH) is the most reactive oxygen radical [29]. It is the most powerful free radical hydroxyl radical found in biological systems. In tissues exposed to radiation, a large part of the energy is absorbed by the water inside the cell and the radiation creates a covalent bond between oxygen and hydrogen, forming hydrogen (H^-) and hydroxyl radical (OH^-).

$$H-O-H \rightarrow H^- + OH^- \text{ (Hydroxyl radical)} \qquad (3)$$

OH^- radicals, which can provide radical formation and participate in a series of reactions, cause strand breaks in DNA by joining the structure of bases in DNA and RNA, which they do by causing a lot of damage to the bases and sugars of DNA. If the damage is very severe, it may not be repaired by cellular protective systems and as a result, mutations and cell death occur [14, 17, 30].

The steady-state concentration of the OH^- radical in vivo is zero because it reacts with every molecule in the living cell, such as DNA, protein, phospholipid, amino acid, and sugar, at or near the place of formation. The high reactivity of the OH^- radical is thought to result from the extraordinary combination of three properties. These properties include high electrophilicity, high thermochemical reactivity, and the ability to form near target molecules [31]. OH^- formation was achieved in living erythrocytes under the effect of adriamycin using the spin trap electron paramagnetic resonance (EPR) technique [32]. Most of the OH^- produced in vivo or in situ was obtained from the decomposition of H_2O_2 [33] by Fe (II) catalysis. Additionally, OH^- can be produced by various sources: sunlight (Joseph JM, Aravindakumar), ultraviolet radiation [34], ionizing irradiation [35], reaction of hypochlorous acid with O2- [36] and sonolysis of water (ultrasound) [37].

The reaction of the OH^- radical can be inhibited by OH^- scavengers such as methanol, ethanol, 1-butanol, mannitol, formate, thiourea, dimethylthiourea, glucose, tris-buffer, or sorbitol [23]. Although OH^- scavengers prevent OH^- from reacting with other molecules, including lipid molecules, they are not always effective. There are several reasons to consider:

1. Reaction of the OH^- radical with a scavenger can create scavenger radicals that can react with other molecules in the system [38].

2. More attention has been paid to the possibility of a metal-mediated mechanism [28]: OH^- produced by the reaction of H_2O_2 with metal ions bound to

macromolecules can react with metal-binding molecules. It has been reported that as a result of the formation of the Fe (II) ion and 2-deoxyreebose complex, the Fe (II) ion that binds to DNA interacts with H_2O_2 to form OH^-, which instantly damages DNA [39]. It was determined that the Fe (III) ion binds to the membrane and then forms free radicals in the binding site. It has been suggested that iron is accepted as the main binding site of the sulfone group with the carboxyl groups of sialic acids to the membrane, the sulfate group of glycolipids and the phosphate head group of glycoproteins and phospholipids [40]. On the other hand, it has also been reported that OH^- scavengers effectively inhibit OH^- formation in the presence of EDTA. Indeed, EDTA allows Fe (II) ions to be removed from these binding sites [41]. Thus, the toxicity of O_2 and H_2O_2 may be due to the presence and distribution of metal ion catalysts to form OH^- in cells.

3.4 Singlet oxygen ($O_2\uparrow\downarrow$)

Singlet oxygen is the name given to the excited form of oxygen; it is a reactive oxygen type with a very high non-radical reactivity. By directly reacting with unsaturated fatty acids, it forms the peroxyl radical and initiates a lipid peroxidation as strong as the hydroxyl radical [10].

4. Free radical sources

Continuously produced SR in the cell and in the environment can be generated by both endogenous and exogenous sources.

Endogenous (Natural) Sources:

- Oxygen catalyzed by the electron transport system during oxygen respiration in mitochondria produces free radicals as a by-product.
- In case of inflammation, cytokines are released and as a result neutrophils and macrophages begin to produce free radicals.
- Free radicals can originate from a variety of sources such as lipid peroxidation, xanthine oxidase, and mitochondrial cytochrome oxidase.
- Free radicals can be produced by smooth muscle cells, platelets, and arachidonic acid metabolism.
- It may occur as a result of electron leaks in the Cytocom p450 system in the endoplasmic reticulum with enzymes such as xanthine oxidase (XO) and nicotinamide adenine dinucleotide phosphate (NADPH) oxidase during autooxidation reactions.
- Stress from mental stress or body fatigue can create free radicals as a toxic by-product. In addition, hormones such as cortisol and catecholamine can cause stress reactions in the body, and these hormones themselves can turn into free radicals.
- Immune system cells can generate ROS and oxy-radicals in response to pathogens.

Exogenous Sources:

- X-rays, UV rays, microwave rays, gamma rays.
- Burning organic materials during cooking.
- Volcanic activities, forest fires.
- Pollutants such as benzene, asbestos, formaldehyde, carbon monoxide, toluene, and ozone.
- Chemicals such as glue, cleaning products, thinner, paint, pesticides, and perfumes.
- Water contaminants such as chloroform and other trihalomethanes.
- Cigarette smoke, exhaust smoke, alcohol and cigarette use can contribute to the production of free radical exogenously [9, 42–45].

5. Damage associated with free radicals

As a result of the damaging effects of free radicals on cells, the cell's plasma and organelle membranes lose their continuity. As a result, sodium and calcium ions enter the cell in addition to water. Morphologically recognized by their pale granular cytoplasm, these cells swell. Over time, this structural defect leads to irreversible changes in the cell0, followed eventually by death [1].

5.1 Lipids

Lipids are the most sensitive biomolecules to the effects of free radicals, and the unsaturated bonds of fatty acids and cholesterol in cell membranes react very easily with free radicals to form peroxidation products. The oxidative breakdown of polyunsaturated fatty acids, known as lipid peroxidation and which is highly harmful, proceeds as a self-sustaining chain reaction [17, 46]. Lipid peroxides, which are an important component of cell membranes, form RS- and ROO- radicals with the presence of transition metals such as Fe and Cu. In this way, Fe and Cu salts increase the rate of lipid peroxidation and consequently reduce the fluidity and permeability of the cell membrane and cause the disruption of membrane integrity [6, 17, 47].

5.2 Lipid peroxidation

Lipid peroxidation is a free radical chain reaction that is comprised of three primary steps: initiation, propagation, and termination. Highly reactive radicals, such as the hydroxyl radical, attack polyunsaturated fatty acids, causing a hydrogen atom to be removed from the methylene (—CH_2—) group and, thus, initiate lipid peroxidation. Polyunsaturated fatty acids are very sensitive to peroxidation, as the number of double bonds in the fatty acid side chain increases, the hydrogen atom cleavage becomes easier [13]. Conjugated dienes will, however, react to one another in the bounds of the membranes or other membrane components such as protein and cholesterol under conditions when O_2 is extremely restricted [48]. The creation of conjugated dienes is followed by changes in the structure of the double bond from cis to trans form, which may facilitate tighter packing of the unsaturated fatty acids, contributing to the development of more rigid domains inside the bilayer of oxidized lipid [49].

When the hydrogen atom leaves the molecule by acquiring an electron, only one electron remains in the carbon of the fatty acid; In order to eliminate the weakening of the C-H bond in the carbon atom adjacent to the double bond, the carbon-centered radical forms the conjugated diene. Conjugated diene reacts with oxygen, causing the lipid peroxyl radical (LOO•); the lipid radical formed in this step is important because it starts a chain reaction by removing the hydrogen atom from another fatty acid. Peroxyl radicals show less reactive properties than •OH; however, they can reach farther regions. Peroxyl radicals can react with each other, attack membrane proteins or break hydrogen atoms from neighboring fatty acid chains, leading to the progression of lipid peroxidation chain reaction. Lipid peroxidation in biological membranes can lead to decreased membrane fluidity and membrane potential, increased permeability to H^+ and other ions, and disruption of organelle or cell integrity [13].

The termination process is the last stage of lipid peroxidation. During this process, LOOs either undergo a reciprocal causal nexus or self-destruct and in this way go on to form non-radical products. Despite their potential to breakdown when exposed to high temperatures or by contact with transitional metal ions, LOOH is a compound which remains stable at physiological temperatures [23]. The formed free radicals (LO•, LOO•) and electrophilic products (e.g. 4-hydroxynonenal) can react with neighboring membrane proteins as well as diffuse with distant molecules such as DNA [13].

Author details

Suzan Onur Yaman[1*] and Adnan Ayhanci[2]

1 Karabuk Unıversıty, Faculty of Health Scıences, Karabuk, Turkey

2 Eskisehir Osmangazı Unıversıty, Faculty of Scıence, Eskısehır, Turkey

*Address all correspondence to: suzanyaman@karabuk.edu.tr

References

[1] Erbiş H. (Prof. Dr. Ergün Erdem). Ratlarda Cerulein Ve Glikodeoksikolik Asit İle Oluşturulan Akut Nekrotizan Pankreatitte Lokal Organ Hasarı Üzerine Tempol'ün Etkisi (Deneysel Çalışma), *Pamukkale Üniversitesi Tıp Fakültesi Genel Cerrahi Anabilim Dalı, Uzmanlık Tezi, Denizli, 2007.*

[2] Meral R, Doğan İS, Kanberoğlu GS. (2012). Fonkiyonel gıda bileşeni olarak antioksidanlar. Iğdır Üni. Fen Bilimleri Enst. Der. / Iğdır Univ. J. Inst. Sci. & Tech. *2(2):45-50.*

[3] Karadeniz F, Koca N. (2005). Serbest radikal oluşum mekanizmaları ve vücuttaki antioksidan savunma sistemleri. *Ankara Üniversitesi, Ziraat Fakültesi, Gıda Mühendisliği Dergisi.*

[4] Çavdar C, Sifil A, Çamsarı T. (1997). Reaktif oksijen partikülleri ve antioksidan savunma. Türk Nefroloji Diyaliz ve Transplantasyon Derg.; *3(4):92-95.*

[5] Codandabany U. (2000). Erythrocyte lipid peroxidation and antioxidants in cigarette smokers. Cell Biochem. Funct.; *18:99-102 .*

[6] Kalaycı H.(2009). Erken Doğum Eyleminde Total Oksidan Seviye, Total Antioksidan Kapasite ve Oksidayif Stres İndeksi. Gaziantep Üniversitesi Tıp Fakültesi, Uzmanlık Tezi, Gaziantep.

[7] Kaya H, Delibaş N, Çapar M, Tahan V, Serteser M, Özkaya MÖ. (1996). Gebelikte ve postpartum erken dönemde serbest radikal oluşumu ve antioksidan enzim düzeyleri. Süleyman Demirel Üniversitesi, Tıp Fakültesi Dergisi; *3(4):67-70.*

[8] Southorn P, Powis G. (1998). Free radical in medicine I. Chemical Nature and Biological Reactions. *J* Mayo Clin Proc.;*63(3):381-388.*

[9] Karabulut H, Gülay MŞ. (2016). Serbest radikaller. *Mehmet Akif Ersoy Ünv.* Sağ.Bil.Enst.Derg.; *4(1):50-59.*

[10] Gülbayzar S. (2006). Yenidoğan Bebeklerde Kord Kanında (Oksidatif Stres Göstergesi Olarak) Malondialdehit. İstanbul Bakırköy Dr. Sadi KONUK Eğitim ve Araştırma Hastanesi Çocuk Sağlığı ve Hastalıkları Kliniği, İstanbul.

[11] Annagür A. (2011). Yenidoğan Sepsisinde Total Antioksidan Seviye, Total Oksidan Seviye ve Serum Paraoksonaz Düzeyleri. Selçuk Üniversitesi, Meram Tıp Fakültesi, Çocuk Sağlığı ve Hastalıkları ABD, Yan Dal Uzmanlık Tezi, Konya.

[12] Memişoğulları R. (2005). Diyabette serbest radikaller ve antioksidanlar. Düzce Tıp Fakültesi Dergisi*; 3:30-39.*

[13] Min B. and Ahn D. U. Mechanism of Lipid Peroxidation in Meat and Meat Products -A Review. *Food Sci. Biotechnol. Vol. 14, No. 1, pp. 152 ~ 163 (2005).*

[14] Brent JA, Rumack HH. (1993). Role of free radicals in toxic hepatic injury 1. *Free Radical Chemistry.J* Toxicol Clin.; *49(4):481-493.*

[15] Pirinççi F. (2014). Açıklanamayan Tekrarlayan Gebelik Kaybı Olgularında M30, M65 Düzeylerinin Sitokin ve Oksidatif Stres Markırları ile Korelasyonu. Harran Üniversitesi, Tıp Fakültesi, Uzmanlık Tezi, Şanlıurfa.

[16] Sezer K, Keskin M. (2014). Serbest oksijen radikallerinin hastalıkların patogenezisindeki rolü. *Fırat Üniversitesi Sağ.Bil.Vet.Derg. ; 28(1):49-56.*

[17] Sucu S. (2012). Ektopik Gebelikte Total Oksidan Seviye, Total Antioksidan Seviye ve Oksidatif Stres Indeksinin Değerlendirilmesi: Prospektif Kontrollü Çalışma. Gaziantep Üniversitesi. Tıp Fakültesi, Uzmanlık Tezi, Gaziantep.

[18] Wetberg AB, A.Weitzman S, P.Clarck E, P.Stossel T. (1985). Effects on antioxidants on antioxidant induce: sister chromatid exchange formation. *J Clin Invest.;75(3):35-7.*

[19] Balagopalakrishna C, Manoharan,PT, Abugo OO, Rifkind JM. (1996). Production of superoxide from hemoglobin-bound oxygen under hypoxic conditions. Biochemistry 35: 6393-6398.

[20] Baron CP, Andersen HJ. (2002). Myoglobin-induced lipid peroxidation. A review. J. Agric. Food Chem. 50: 3887-3897.

[21] Sanders-Stephen A, Eisenthal R, Harrison R. (1997). NADH oxidase activity of human xanthine oxidoreductase: Generation of superoxide anion. Eur. J. Biochem. 245: 541-548.

[22] Thomas EL, Learn DB, Jefferson MM, Weatherred W. (1988). Superoxide-dependent oxidation of extracellular reducing agents by isolated neutrophils. J. Biol. Chem. 263: 2178-2186.

[23] Kanner J. (1994). Oxidative processes in meat and meat products: Quality implications. Meat Sci. 36: 169-189.

[24] Giulivi C, Hochstein P, Davies KJA. (1994). Hydrogen peroxide production by red blood cells. Free Rad. Biol. Med. 16: 123129.

[25] Chan WKM, Faustman C, Yin M, Decker EA. (1997). Lipid peroxidation induced by oxymyoglobin and metmyoglobin with involvement of H_2O_2 and superoxide anion. Meat Sci. 46: 181190.

[26] Harel S, Kanner J. (1985). Hydrogen peroxide generation in ground muscle tissues. J. Agric. Food Chem. 33: 1186-1188.

[27] Halliwell B, Gutteridge JMC. (1985). Oxygen radicals and the nervous system. Trends Neurosci. 79: 22-26.

[28] Lamba OP, Borchman D, Garner WH. (1994). Spectral characterization of lipid peroxidation in rabbit lens membranes induced by hydrogen peroxide in the presence of Fe^{2+}/Fe^{3+} cations: A site specific catalyzed oxidation. Free Rad. Biol. Med. 16: 591-601.

[29] Halliwell B, Gutteridge JMC. (1999). Free radicals in biology and medicine. 3rd ed. Oxford University Press, New York, NY,USA. pp. 36-350.

[30] Dizdaroğlu M. (1993). Chemical determination of free radical-induced damage to DNA. Free Radic Biol Med.*; 61(3):225-242.*

[31] Pryor WA. (1988). Why is the hydroxyl radical the only radical that commonly adds to DNA. Hypothesis: It has a rare combination of high electrophilicity, high thermochemical reactivity, and a mode of production that can occur near DNA. Free. Rad. Biol. Med. 4: 219-223.

[32] Bannister JV, Thornalley PJ. (1983). The production of hydroxyl radicals by Adriamycin in red blood cells. FEBS Lett. 157: 170-172.

[33] Joseph JM, Aravindakumar CT. (2000). Determination of rate constants for the reactions of hydroxyl radicals with some purines and pyrimidines using sunlight. J. Biochem. Biophys. Methods 42: 115-124.

[34] Floyd RA, West MS, Eneff KI, Hogsett WE, Tingey DT. (1988). Hydroxyl free radical mediated formation of 8-hydroxyguanine in isolated DNA. Arch. Biochem. Biophys. 262: 266272.

[35] Swallow AJ. (1984). Fundamental radiation chemistry of food

components. pp. 165-177. In: Recent Advances in the Chemistry of Meat. Bailey AJ (ed). Royal Soc. Chem. Burlington House, London, UK.

[36] Folkes LK, Candeias LP, Wardman P. (1995). Kinetics and Mechanisms of hypochlorous acid reactions. Arch. Biochem. Biophys. 323: 120-126.

[37] Henglein A, Kormann C. (1985). Scavenging of OH radicals produced in the sonolysis of water. Int. J. Radiat. Biol. Relat. Stud. Phys. Chem. Med. 48: 251-258.

[38] Miller GG, Raleigh, JA. (1983). Action of some hydroxyl radical scavengers on radiation-induced haemolysis. Int. J. Radiat. Biol. Relat. Stud. Phys. Chem. Med. 43: 411-419.

[39] Auroma OI, Grootveld M, Halliwell B. (1987). The role of iron in ascorbate-dependent deoxyribose degradation. Evidence consistent with a site-specific hydroxyl radical generation caused by iron ions bound to the deoxyribose molecule. J. Inorg. Biochem. 29: 289-299.

[40] Baysal E, Sullivan SG., Stern A. (1989). Binding of iron to human red blood cell membranes. Free Rad. Res. Comms. 8: 55-59.

[41] Gutteridge JM. (1984). Reactivity of hydroxyl and hydroxyl-like radicals discriminated by release of thiobarbituric acid-reactive material from deoxy sugars, nucleosides, and benzoate. Biochem. J. 224: 761-767.

[42] Bagchi K, Puri S. (1998). Free radicals and antioxidants in health and disease. East Mediterr Health J. *4(2):350-360.*

[43] Pham-Huy LA, He H, Pham-Huy C. (2008). Free radicals, antioxidants in disease and health. *Int J Biomed Sci. 4(2):89-96.*

[44] Sen S, Chakraborty R, Seidhar C, Reddy YSR, De B. (2010). Free radicals, antioxidants, diseases and phytomedicines: current status and future prospect. Int J Pharm Sci Res. *3(1), 91-100.*

[45] Yoshıkawa T, Ueda S, Natto Y, Takahassı S, Oyamada H, Morıta Y, Yoneta T, Kondo M. (1996). Role of oxygen-derived free radicals on gastric mucosal injury induced by ischemia-reperfusion. Saudi J Gastroenterol. *2(1):19-28.*

[46] Akpoyraz M, Durak İ. (1995). Serbest radikallerin biyolojik ertkileri. Ankara Tıp Mecmuası (The Journal Of The Faculty Of Medicıne) Vol.; *48:253-262.*

[47] Sürücü M. (2014). Gestasyonel Diabetli Hastalarda Total Oksidatif Seviye Total Antioksidan Seviye ve Oksidatif Stres İndeksinin Değerlendirilmesi. Gaziantep Üniversitesi Tıp Fakültesi, Kadın Hastalıkları ve Doğum Anabilim Dalı, Uzmanlık Tezi, Gaziantep.

[48] Frank H, Thiel D, MacLeod J. (1989). Mass spectrometric detection of cross-linked fatty acids formed during radical-induced lesion of lipid membranes. Biochem. J. 260: 873-878.

[49] Coolbear KP, Keough KM. (1983). Lipid peroxidation and gel to liquid-crystalline transition temperatures of synthetic polyunsaturated mixed-acid phosphatidylcholines. Biochim. Biophys. Acta. 732: 531-540.

Chapter 2

Lipid Peroxidation: A Signaling Mechanism in Diagnosis of Diseases

Kalpana Sabanna Patil and Raju Ratan Wadekar

Abstract

Quantification of reactive oxygen species, is perplexing either *in vivo* or *in vitro* due to their short half-lives. Consequently, to define the magnitude of oxidative stress, the more stable oxidation products can be measured in biological samples. The oxidative stress leads to the lipid peroxidation that involves the initiation, termination and propagation of lipid radicals, wherein, the process involves the oxygen uptake, rearrangement of the double bonds in unsaturated lipids, that leads to polyunsaturated fatty acid deterioration. Subsequently, the toxic signaling end products are considered as biomarkers of free radicals that act both as signaling molecules and as cytotoxic products cause covalent alteration of lipid peroxidation products. The use of validated signaling mechanism (s) of Lipid peroxidation and products derived thereof exhibits its use clinical practice and basic clinical research as well as in clinical practice has become common place, and their presence as endpoints in clinical trials is now broadly accepted. This knowledge can be used to diagnose disease earlier, or to prevent it before it starts. The signaling markers can be used to excel the effectiveness of the prevailing medicines and to improve the new medicines.

Keywords: lipid peroxidation, isoprostanes, malondialdehyde, Alzheimer's disease, oxidative stress

1. Introduction

Lipids are of two types: Polar and Non-polar. The polar lipids (Triglycerides), store in various cells but especially in adipose (fat) tissue, are usually the main source of energy for mammals. Polar lipids are underlying segments of cell layers, where in it contributesfor thedevelopment of permeability barrier of cells and sub-cellular organelles in the form of a lipid bilayer [1]. The glycerol-based phospholipid is the significant type of membranous lipid bilayer and it is evidenced by the element that membrane lipids may regulates the biologicalfunctions of a membrane organelle by amending its biophysical characteristics, such as the divergence and absorptivity [2]. Lipids and its metabolite productsfacilitates a key ingredient in understanding the biology and serve as a signaling biomolecules in the diagnosis of diseases [3]. However, theleading enzymes that generate as lipid-signaling biomarkers are lipoxygenase, that intervene hydroperoxyeicosatetraenoic acids (HPETEs), lipoxins, leukotrienes, or hepoxilins biosynthesis after oxidation of fatty acids/arachidonic acid (AA), cyclooxygenase that yields prostaglandins, and cytochrome P-450 (CYP) that produces epoxyeicosatrienoic acids, leukotoxins, thromboxane, or prostacyclin respectively [4, 5]. The signaling lipid biomarkers recruitsvia stimulation of a variety of receptors, including nuclear and G protein-coupled receptors. Moreover, several othertypes of lipidmetabolites have been recognized as potent intracellular signal transduction molecules viz; i) diacylglycerol (DAG) and inositol phosphates (IPs) were derived from the phosphatidylinositol phosphates. DAG is a transcription nuclear factor-kB (NF-kB) which promotes cell survival and proliferation and also a physiological activator of protein kinase C [6, 7] and a small G protein [8]. On the other hand, IPs (lipid derived metabolites) are anextremelystimulatingthat intricate in signal transduction, results in activation of mTORand Akt [9], and calcium homeostasis [10, 11]; ii) Sphingolipid derived from ceramide (sphingosine-1-phosphate), is aeffective messenger molecule engaged in proliferation, adhesion, migration and alsoregulates calcium mobilization at molecular and cellular level of the organism [12–14]; iii) oxidative stress induced fatty acid derived eicosanoid and prostaglandins involved in inflammation [15, 16] and immunity [17]; iv) phosphotidylserine, (a phospholipid) that plays crucial role in a number of signaling pathways, includes fusogenic proteins, kinases and small GTPases [18]; v) the sex and growth hormones such as testosterone, progesterone, estrogen and cortisol that monitored a host body activities such as reproduction, blood pressure metabolism, inflammation, oxidative stress response etc. [19].

Molecular mechanism of lipid damage: The process of lipid peroxidation (LPO), is the resultant of oxidative stress and free radical production. Specifically, reactive oxygen species (ROS) attack polyunsaturated fatty acids (PUFAs) of cellular membranes and leads tothe insult of functional and/or structural integrity of cell membranes, subsequentlyproducing4-hydroxy-2-noneal (HNE), malondialdehyde (MDA) and acrolein (a group of α, β-unsaturated highly reactive aldehyde) [20, 21]. Therefore, these strong reactive aldehydes are significantlydiffusive, able to attack and form covalent linkages with auxiliary cellular constituents. Moreover, the lipid peroxidation process continues asself-propagation followed by initiation of chain reactions and termination either with complete substrate utilization or through interaction with antioxidants such as tocopherol (Vitamin E). Neuroprostanes (neuroPs), isoprostanes (IsoPs) are the additional LPO products of arachidonic acid and docosahexaenoic acid (DHA), that are quantified in the biological fluids to diagnose the severity of the disease. Furthermore, the cyclized fatty acids proliferate further and metabolize the cellular membrane components, mainly lipids and proteins, and propagates the other LPO products in the body fluids [22].

Quantification of reactive oxygen species, is perplexing either *in vivo* or *in vitro* due to their short half-lives. Consequently, to define the magnitude of oxidative stress, the more stable oxidation products can be measured in biological samples. The oxidative stress leads to the lipid peroxidation that involves the initiation, termination and propagation of lipid radicals, wherein, the process involves the oxygen uptake, rearrangement of the double bonds in unsaturated lipids, that leads to polyunsaturated fatty acid deterioration. Subsequently, the toxic signaling end products are considered as biomarkers of free radicals that act both as signaling molecules and as cytotoxic products cause covalent alteration of lipid peroxidation products [23]. In respect of their oxidative-induced damage properties, these compounds are considered as disease mediators in the pathophysiology of many neurodegenerative diseases (NDs), including Alzheimer's disease (AD), Parkinson's disease (PD), and amyotrophic lateral sclerosis (ALS), Diabetes, Atherosclerosis, Chronic inflammation, Asthma and liver injury that serve as potential biomarkers in the signaling mechanism in diagnosis of diseases [24]. Thus, it is necessary to understand the oxidative deterioration of lipids in a sequential five-step procedure

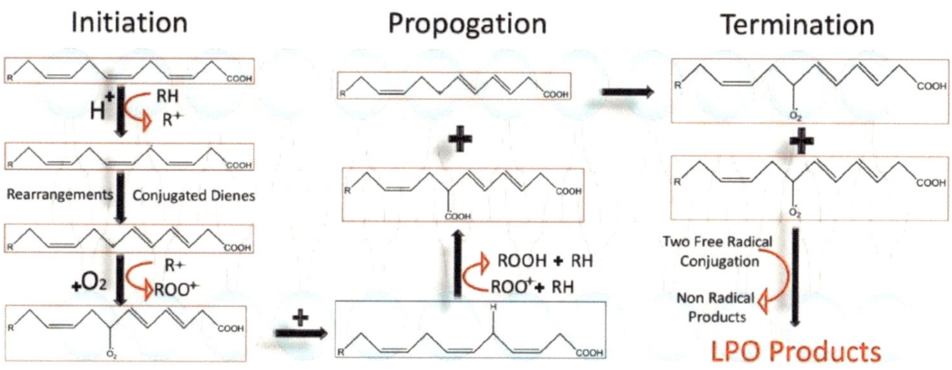

Figure 1.
Oxidative deterioration of lipids.

in which oxidants, either radical or non-radical species, attack lipids containing C-C double bonds [25, 26]. On the contrary of enzyme-based lipid metabolism, lipid peroxidation does follow a non-enzymatic process that continues in ahystericalmode: Initiation, propagation and termination (**Figure 1**) [27].

2. Mechanism of action of lipid peroxidation (LPO)

The process of LPO on membrane influences discrete functions from the increased rigidity of membrane, reduced action of membrane-confined enzymes, impairment of membrane receptors and modified permeability of the cell membrane. Similar to phospholipid impairment, radicals can also directly attack lipid-protein and membrane proteins mediate as well as protein–protein interconnection, subsequently affect the membrane integrity [28]. LPO products persuade such a loss of membrane integrity that ultimately leads to unadorned cytotoxicity, and could result in unrestrained cellular growth or even apoptosis. Rationally, the perturbation of the above-mentioned functions ensued by polyunsaturated fatty acids, along with the resultant metabolites and protein insults modifies the neuronal homeostasis, and leads to the multi-organ organ dysfunction [29–31].

3. Lipid peroxidation (LPO) products as biomarkers in neurodegenerative disorders

LPO products are significantly associated to the development of Alzheimer's diseases (AD); and hence, they are studied as potential disease signaling biomarkers in neurodegenerative disorders. LPO products such as MDA, IsoPs, TBARS, and fluorescent lipofuscin-like pigments (LPF) extensively studied and found in different human samples (plasma, serum and urine) of the patients suffering from neurodegenerative disorders. (Summarized in **Table 1**).

Histopathological studies revealed a co-localization of lipid peroxidation products and β-amyloidplaques in the brain of the AD. Also, the studiy evident for the presence of fatty acids in AD brain lesions produced a neurotoxic effect in cell culture increasing oxidative stress [41]. Since the brain contains high lipid content and high oxygen consumption, lipid peroxidation seems to play fundamental role in AD early detection. Similarly, IsoPsand its isomers produced via diverse actions that, are encountered as marginal oxidation products of the arachidonic acid [42]. Whereas, neuroPsenriched in the neuronal tissue and vital component

Sr. no	Biomarkers	Biological sample	Analytical technique	Results	Reference
01	8-Isoprostane	Urine	EIA	AD <DrD* AD+DM < DrD	[32]
02	Isoprostanes oxidized LDL	Urine	ELISA	Not differences between groups	[33]
03	8-Isoprotanes	Serum	ELISA	Non-frail AD < Pre-frail AD*	[34]
04	Isoprostanes	Urine	EIA	Non-frail AD < frail AD*	[35]
05	Isoprostanes, Neuroprostanes, dihomo-isoprostanes	Urine	RIA	AD + placebo > AD treatment	[36]
06	MDA	Urine	GC–MS	Not differences between groups after the treatment	[37]
07	MDA	Urine	UPLC-MS/MS	Significant differences in groups	[38]
08	MDA	Plasma	HPLC-fluoresce	aMCI converted >aMCI stable	[39]
09	MDA	Blood	HPLC-MS	MDA blood levels do not correlated with different cognitive tests	[40]

Table 1.
Signaling mechanism of lipid peroxidation in biological samples of patients.

of the nervous tissues, awfully susceptible to oxidation [43]. Thus, the quantitative estimation of neuroPsaffords a significantsource of oxidative neuronal impairment-corresponding to IsoPs [44].

Malondialdehyde (MDA) a signaling molecule of LPO has ability to interact with micro-macromolecules such as nucleic acid bases, developingdivergent adducts, and can also react with proteins in a synergistic and covalent manner, subsequently, leads to the stimulation of strong immune responses and exhibits pro-fibrogenic and pro-inflammatory properties/mediators such as interleukins, cytokines etc. Furthermore, accumulation of MDA modifies membrane integrity by inducing increased intra and extracellular permeability and damage the fluidity of membrane lipid bilayer. Being a most mutagenic, MDA is capable of reacting with deoxyadenosine in DNA and deoxyguanosine, thus generating mutagenic DNA adducts [21, 31].

As the consequence of peroxidation of PUFAs (linoleic and arachidonic acid), Hudroxy-2-nonrenal (HNE) are formed, since they are the most abundant in fatty acids. The HNE, specifically, bind to amino acids mainly: cysteine, histidine and lysine proteinaceous residue addition by either the amino and thiol groups. The conjugates of protein residue and HNE, leads to the impairment of the normal protein function as well as structure, and also HNE exhibits reactivity with vital nucleic acids, lipids, signaling biomolecules and vitamins. Documented reports, suggests that, the HNE accumulates in extremely low concentration (10 μM), in response to oxidative stress and induces cytotoxicity and selective suppression of inducible and basal NF-kB factors. Therefore, increased levels of HNE results in Ca^{2+} homeostasisimbalance, disruption of glutamate transport, membrane impairment, microtubule function, and cellular death via the activation of caspase pathways [28, 45].

Threonine metabolite product, acrolein generated by the bio activation of phagocytes and cyclophosphamide. Wherein, acrolein targets histidyl, lysyl and cysteinyl residue of protein side chain as well as reactswith nucleophilic sites in DNA, that results in DNA and protein adducts and, thus, initiates cytotoxicity specifically related to its ability to reduce glutathione [46]. Docosahexaenoic acid (DHA) enriched in neurons, and is a vital compound of the nervous tissue. It is a vital compound of the nervous tissue and enriched in neurons and extremely-susceptible to oxidation. DHA on oxidative stress, leads to the production of Neuroprostanes (F4-isoprostanes). In a biological aspect, neuroPsillustrates anti-inflammatory properties by inhibiting proteasome concentrated in the neurons membrane [45]. Nevertheless, the central nervous system (CNS) is one of the major targets of the LPO and proneto chain reactions induced by ROS, which eventually result in LPO products [47]. The role of LPO quantification in the pathogenesis of NDs is significant and extremely importance for the early detection of neurodegenerative disorders [45].

The most frequently exploited LPO products such as lysine residues and unsaturated aldehydes, including HNE and aracolein [48]. Several research studies have been probing the LPO products and disease state interrelation, and its application as possible biomarkers in order to assess prognosis and establish early detection of the disease [49]. Among the above-mentioned potential biomarkers, IsoPssignifies the most reliable and robust outcomes. Moreover, the IsoPs accurately process and assessed the oxidant stress *in vivo*via quantification of plasma and urinary sample. Also, *in situ* phospholipids composed of IsoPs that locates the free radical production and release from the cellular membrane via phospholipases in the plasma. IsoPs detected and quantified in a plethora of biological fluids including plasma cerebrospinal fluid, exhaled breath condensate, urine and bile [50]. On the other hand, neuroPs are a fundamental component of the nervous tissue, enriched in the neuronal tissue and extremely susceptible to oxidation [51]. Thus, the quantification of neuroPs provides a signaling biomarker of oxidative neuronal damage compare to IsoPs quantification. In addition, the quantity of neuroPs produced

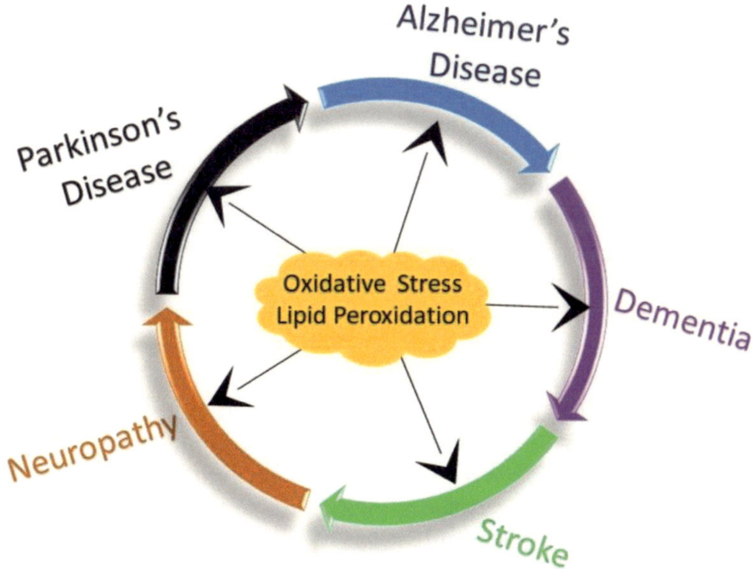

Figure 2.
Lipid peroxidation (LPO) products as biomarkers in neurodegenerative disorders.

from DHA surpass that of IsoPs from arachidonic acid by 3.4 folds. NeuroPsare elevated in the cerebrospinal fluid and brain tissue in ND, such as Parkinson and Alzheimer's disease. Hence, quantification of neuroPs levels isa vital tool in evaluating brain oxidative impairment [52]. Whereas, crosslinking is a major factor in the development of pathology due to the promotion of intramolecular or intermolecular DNA and protein cross-linking, which results in intense change in the biochemical properties of various biomolecules (**Figure 1**). This articulated process is assumed to be a channel of interrelation chain reactions with covalent nucleophilic compounds. Also, the translated and interconnected experimental indicators with precise altered proteins in the CNS exhibited those definite cellular amendments are in concomitant with pathophysiology of Neurodegenerative diseases. Thus, the revival of scientific data affords a comprehensive knowledge in the advancement and employment of LPO products as potential biomarkers in the early diagnosis of the disease, alteredbiological processes, revealing potential active sites to target disease progression (**Figure 2**).

4. Lipid peroxidation metabolites as influential signaling biomarkers in asthma and airway inflammation

Oxidative stress at molecular and cellular level can have many detrimental effects on airway function, including airway smooth muscle contraction, induction of airway hyper responsiveness, mucus hypersecretion, vascular exudation and shedding of epithelial cells. Furthermore, ROS can induce cytokine and chemokine production through induction of the oxidative stress-sensitive transcription of nuclear factor–kB in bronchial epithelial cells [53]. Recently discovered series of bioactive prostaglandin (PF)F2-like compoundswere produced independently of the cyclooxygenase enzymes via the peroxidation of arachidonic acid, catalyzed by free radicals. The pathway leads to formation of 64 isomeric structures, of which 8-iso-PGF2α is most well characterized. Evidence suggests that 8-iso-PGF 2α may act in part through the vascular thromboxane A2/PGH2 (TP) receptor [54]. 8-iso-PGF2α has been found to elicit airway hyper-responsiveness in isolated perfused mouse lungs, and cause airway obstruction and air plasma exudation in guinea pigs in *vivo* [55]. These experimental findings offerassumption about the contribution of Isoprostanes to the airway narrowing that is characteristic of asthma and in addition to being reliable signaling marker of lipid peroxidation, Isoprostane may prove to have an important biological role in the pathological of asthma. A significant elevation of reactive oxygen species, MDA formation (A product of Lipid peroxidation) and Isoprotane was estimated in the broncho-alveolar lavage (BAL) fluidwithin 24 hrs of allergen-induced asthma. This clearly indicates, Isoprotane is produce as a biomarker in respiratory tract tissues that leads to the late observed physical symptoms in allergen-induced asthma [56].

A recent study demonstrated that concentrations of exhaled ethane were increased in patients with more severe bronchoconstriction (forced expiratory volume in one second (FEV1) <60%), compared with less-constricted patients (FEV1 > 60%) and provides evidence that lipid peroxidation is related to asthma severity. These relationships between markers of oxidative stress and disease severity suggest that such tests may indicate the clinical status of asthma patients [57]. The increased level of 8-sio-PGF2α concentrations has been observed in chronic obstructive pulmonary diseases and asthma [58]. The discovery of Isoprostane has generated attention, as they provide a reliable index of oxidative stress *in vivo*. Isoprostane are structurally stable, are produced *in vivo* and are present in relatively high concentrations [59]. Traceable levels of F2-isoprostanes can be found in all normal animal and

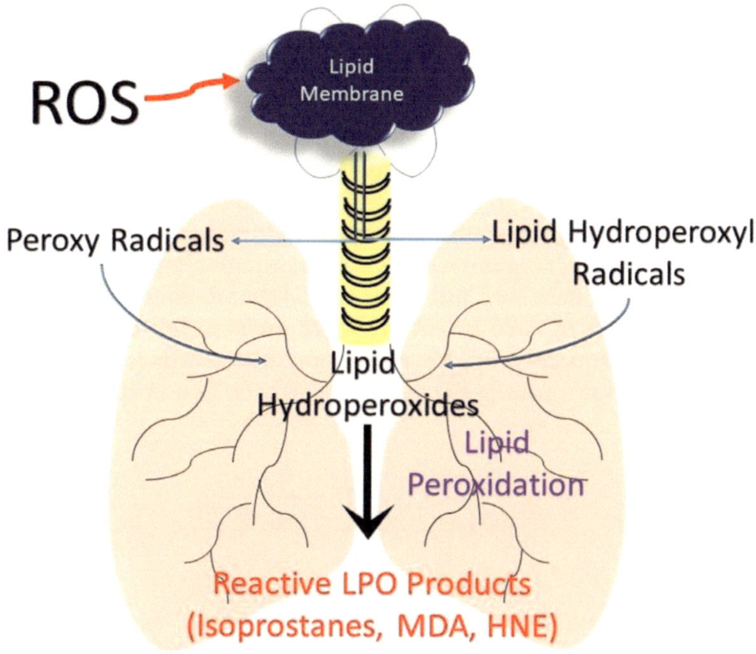

Figure 3.
Lipid peroxidation metabolites as influential signaling biomarkers in lung diseases.

human biological fluids (including plasma, urine, bile, gastric juice, synovial fluid and cerebrospinal fluid)), and esterified in normal animal tissues. Thus, they overcome many of the methodological problems associated with other signaling markers. Determination of 8-iso-PGF2αas a marker of oxidative stress, of carbon tetrachloride (CCl_4)-induced lipid peroxidation has been shown to be 20 times more sensitive than measurement of Thiobarbituric acid reactive substances (TBARS). Thus, the reliability of isoprostanes as in vivo markers of lipid peroxidation makes them an extremely valuable signaling biomarker for defining the potential of antioxidant agents (Vitamin C, E and β-carotene) in humans [60]. A significant amount of elevated ethane produced following lipid peroxidation has been observed in plasma and breath condensate of asthmatics as a biomarker indicator of acute airway inflammatory diseases. Moreover, measurement of auto-antibodies directed against oxidative modifications of low density lipids (LDL) is a recently developed technique that provides an in vivo marker of lipid peroxidation. Enzymes-linked immunosorbent assays are available in kit form, providing a quick and simple methodology. Thus measurement of isoprostanes in breath condensate should provide useful information concerning the degree of oxidant stress and success of antioxidant therapy in asthma (**Figure 3**).

5. Lipid peroxidation: a signaling mechanism in diagnosis of liver injury

Oxidative stress is one of the mechanisms involved in the pathogenesis of drug-induced reactive oxygen species which lead to the depletion of intracellular antioxidants, causing an imbalance in the redox status of the hepatic cells [61]. Rapid, extensive lipid peroxidation of the membrane structural lipids due to oxidative stress mechanism involved in the pathogenesis of drug-induced had seen proposed as the basis of drug-induced hepatocellular toxicity. The most of the xenobiotics such as Acetaminophen, Isoniazid and Rifampicin are well-known to induced hepatic damage directly or indirectly via lipid peroxidation [62]. However, peroxy radical attribute to lipid peroxidation, thus known for the destabilization and disintegration of the cell membrane, that further causes arteriosclerosis, hepatic and kidney damage. The increased serum markers such as MDA formation are of diagnostic importance of hepatic injury because they are released due to the damage of hepatocytes and consequently participate in endogenous enzymatic antioxidant system imbalance [63]. CCl4-ehanced lipid peroxidation has been observed in liver tissue homogenates, isolated hepatocytes and in vivo, and this has been associated with changes in endoplasmic reticular enzyme activity, in vivo fatty acid export and protein synthesis [64]. CCL4 metabolism enhances production of malondialdehyde in vitro and increase ethane production and lethality in vivo (**Figure 4**). Consequently, lipid peroxidation initiated by free radical reactions and unchecked by compromised cellular defenses, provides a possible link between

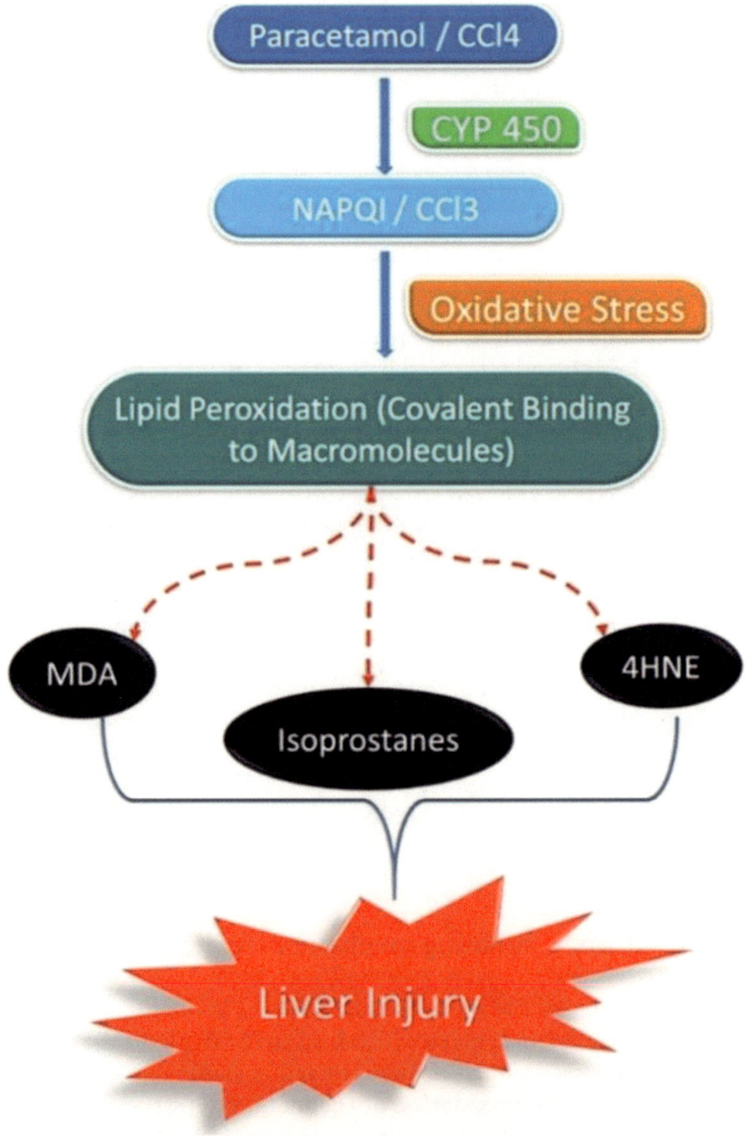

Figure 4.
Lipid peroxidation: a signaling mechanism in diagnosis of liver injury.

ethanol metabolism and associated liver disease [65]. The lipid peroxide content of liver is elevated by both short and long-term ethanol exposure and an enhanced rate of lipid peroxide formation following ingestion has been ascertained by MDA production, diene conjugate formation and *in vivo* ethane and pentane exhalation. Lipid peroxidation might merely be a sign of oxidative processes which occur after reduced glutathione is depleted concomitant with free radical attack on cellular protein and nucleic acid. The elevated MDA formation as a product of lipid peroxidation in drug-induced liver damage provide a significant biological signaling marker for the early detection or diagnosis of liver injury.

6. Conclusion

The oxidative stress inducing compounds mediates metabolic process of lipids mainly via peroxidation that leads to the production of macromolecules such as Isoprostane, MDA, 4-HNE in the biological fluids. Moreover, the aldehyde like molecules produced via lipid peroxidation targets and modifies proteins and DNA substantially at macromolecule level. Furthermore, MDA and 4-HNE known to promote cross linking of protein/DNA reactions that significantly alleviates and alters the biomarkers biochemical property, thereby develops a clinical symptomatic states. The use of validated signaling mechanism (s) of Lipid peroxidation and products derived thereof in basic and clinical research as well as in clinical practice has become common place, and their presence as endpoints in clinical trials is now broadly accepted. This knowledge can be used to diagnose disease earlier, or to prevent it before it starts. The signaling markers can be used to improve the efficacy and safety of existing medicines and to develop new medicines.

Conflict of interest

The authors declare no conflict of interest.

Author details

Kalpana Sabanna Patil[1*] and Raju Ratan Wadekar[2]

1 K.L.E.'s College of Pharmacy, Belagavi, Karnataka, India

2 Department of Pharmacognosy and Phytochemistry, Shri Vile Parle Kelavani Mandal's Institute of Pharmacy, Dhule, Maharashtra, India

*Address all correspondence to: kalpatil@yahoo.com

References

[1] Frunbeck G, Gomez-Ambrosi J, Muruzabal FJ and Burrell MA. The adipocyte: a model for integration of endocrine and metabolic signaling in energy metabolism regulation. The American Journal of Physiology: Endocrinology and Metabolism. 2001; 280:827-847.

[2] Frayn KN. Regulation of fatty acid delivery in vivo. Advances in Experimental Medicine and Biology. 1998; 441: 171-179.

[3] Vance E and Vance JE. Biochemistry: Biochemistry of Lipids, Lipoproteins and Membranes, 4th ed. 2002. p. : 144-155.

[4] Massey KA and Nicolau A. Lipidomics of polyunsaturated fatty-acid-derived oxygenated metabolites. Biochemical Society Transactions. 2011; 39: 1240-1246.

[5] Massey KA and Nicolaou A. Lipidomics of oxidized polyunsaturated fatty acids. Free Radical Biology and Medicine. 2013; 59:45-55.

[6] Jornayvaz FR and Shulman GJ. Diacylglycerol activation of protein kinase Cε and hepatic insulin resistance. Cell Metabolism. 2012; 15:574-584.

[7] Giorgi C, Agnoletto C, Baldini*et al*. Redox control of protein kinase C: cell and disease specific aspects. Antioxidants and Redox Signaling. 2010; 13: 1051-1085.

[8] Yang C and Kazanietz MG. Chimaerins: GAPs that bridge diacylglycerolsignaling and the small G-protein Rac. Biochemical Journal. 2007; 403:1-12.

[9] Baumann J, Sevinsky C and Conklin DS. Lipid biology of breast cancer. Biochimica Biophysica Acta. 2013; 1831:1509-1517.

[10] Fisher SK, Novak JE and Agranoff BW. Inositol and higher inositol phosphates in neural tissues: homeostasis, metabolism and functional significance. Journal of Neurochemistry. 2002; 82:736-754.

[11] Conway SJ and Miller GJ. Biology-enabling inositol phosphates, phosphatidylinositol phosphates and derivatives. Natural Product Reports. 2007; 24:687-707.

[12] Takuwa Y, Okamoto Y, Yoshioka K and Takuwa N. Sphingosine-1-phosphate signaling in physiology and diseases. Biofactor. 2012; 38:329-337.

[13] Mattson MP. Membrane Lipid signaling in aging and age related disease. Elsevier. 2003;

[14] Hannun YA and Obeid LM. Principles of bioactive lipid signaling lessons from sphingolipids. Nature Reviews Molecular Cell Biology. 2008; 9:139-150.

[15] Aoki T and Narumiya S. Prostaglandins and chronic inflammation. Trends in Pharmacological Sciences. 2012; 33:304-311.

[16] Tang EHC, Libby P, Vanhoutte PM and Xu A. Anti-inflammatory therapy by activation of prostaglandins EP4 receptor in cardiovascular and other inflammatory diseases. Journal of Cardiovascular Pharmacology. 2012; 59:116-123.

[17] Kalinski P. Regulation of immune responses by prostaglandins E_2. Journal of Immunology. 2012; 188:21-28.

[18] Kay JG and Grinstein S. Phosphatidylserine-mediated cellular signaling. Advances in Experimental Medicine and Biology. 2013; 991:177-193.

[19] Pluchino N, Russo M, Santoro P et al. Steroid hormones and BDNF. Neuroscience 2013; 239:271-279.

[20] Signorini C, Felice DC, Durand T et al. Isoprostanes and 4-hydroxy-2-nonenal: markers or mediators of diseases? Focus on Rettsyndrome as a model of autism spectrum disorder Oxidative Medicinal Cell Longevity. 2013; 2013:343-824.

[21] Erejuwa OO, Sulaiman SA, Wahab MS. Evidence in support of potential applications of lipid peroxidation products in cancer treatment. Oxidative Medicinal Cell Longevity. 2013; 2013:931-251.

[22] Zarkovic K. 4-hydroxynenal and neurodegenerative diseases. Molecular Aspects of Medicine. 2003; 24:293-303.

[23] Sultana R, Perluigi M, Allan BD. Lipid peroxidation triggers neurodegeneration: a redox proteomics view into the Alzheimer disease brain. Free Radical Biological Medicine. 2013; 3:157-169

[24] Barrera G, Gentile F, Pizzimenti S, et al. Mitochondrial Dysfunction in Cancer and Neurodegerataive Diseases: Spotlight on Fatty Acid Oxidation and Lipid Peroxidation Products. Antioxidants (Basel). 2016; 5:1

[25] Perluigi M, Coccia R, Butterfield DA. 4-Hydroxy-2-nonenal, a reactive product of lipid peroxidation, and neurodegenerative diseases: a toxic combination illuminated by redox proteomics studies. Antioxidant Redox Signal. 2012; 17:1590-1609.

[26] Ayala A, Munoz MF, Arguelles S. Lipid peroxidation: production, metabolism, and signaling mechanisms of malondialdehyde and 4-hydroxy-2-nonenal. Oxidative Medicinal Cell Longevity. 2014; 2014:360-438.

[27] Bradley-Whitman MA, Lovell MA. Biomarkers of lipid peroxidation in Alzheimer diseases (AD): an update. Archive Toxicology. 2015; 89:1035-1044.

[28] Shoeb M, Ansari NH, Srivastava S., et al. 4-hydroxynonenal in the pathogenesis and progression of human diseases. Current Medicinal Chemistry. 2014; 21: 230-237.

[29] Yadav UC, Ramana KV. Regulation of NF-kappaB-induced inflammatory signaling by lipid peroxidation-derived aldehydes. Oxidative Cell Longevity. 2013; 2013:1-13

[30] Fritz KS, Petersen DR. An overview of the chemistry and biology of reactive aldehydes. Free Radical Biology medicine. 2013; 59:85-91.

[31] Gueraud F, Atalay M, Bresgen N, et al. Chemistry and biochemistry of lipid peroxidation products. Free Radical Research. 2010; 44:1098-1124.

[32] Hatanaka H, Hanyu H, Fukasava R, et al. Peripheral oxidative stress markers in diabetes-related dementia. Geriatric Gerontology International. 2016; 16:1312-1318.

[33] Scheff SW, Ansari MA, Mufson EJ. Oxidative stress and hippocampal synaptic protein levels in early cognitively intact individuals with Alzheimer's disease pathology. Neurobiological aging. 2016; 42:1-12.

[34] Rubio Perez JM , et al. Effect of an antioxidant beverage on biomarkers of oxidative stress in Alzheimer's patients. European Journal of Nutrition. 2016; 55:2105-2116.

[35] Namioka N, Hanyu H, Hirose D, Hatanaka H, Sato T, Shimizu S. Oxidative stress and inflammatory are associated with physical frailty in patients with Alzheimer's disease. Geriatric Gerontology International. 2017; 17: 913-918.

[36] Freund-Levi Y, Vedin I, Hjorth E, et al. Effects of supplementation with

omega-3-fatty acids on oxidative stress and inflammation in patients with Alzheimer's disease: the omega AD study. Journal of Alzheimer's Disease. 2014; 42:823-831.

[37] Shinto L, Quinn J, Montine T, et al. A randomized placebo-controlled pilot trial of omega-3-fatty acids and alpha lipoic acid in Alzheimer's disease. Journal of Alzheimer's Disease. 2014; 38:111-120.

[38] Rommer PS, Fuchs D, Leblhuber F, et al. Lowered levels of carbonyl proteins after vitamin B supplementation in patients with mild cognitive impairment and Alzheimer's disease. Neurodegenerative Disease. 2016; 16:284-289.

[39] Yuan L, Liu J, Ma W, et al. Dietary pattern and antioxidants in plasma and erythrocyte in patients with mild cognitive impairment from China. Nutrition. 2016; 32:193-198.

[40] Gubandru M, Margina D, Tsitsimpikou C, et al. Alzheimer's patients showed different patterns for oxidative stress and inflammation markers. Food Chemistry and Toxicology. 2013; 61:2019-2214.

[41] Benseny-cases N, Klementieva O, Cotte M, et al. Microspectroscopy reveals co-localization of lipid oxidation and amyloid plaques in human Alzheimer's disease brains. Analytical Chemistry. 2014; 86:12047-12054.

[42] Zarrouk A, Nury T, Riedinger JM, et al. Effect of docosahexaenoic acid (attenuation or amplification) on C22:0-, C24:0-, and C26:0-induced mitochondrial dysfunctions and oxidative stress on human neuronal SK-N-BE cells. Journal of Nutrition Health and Aging. 2015; 19:198-205.

[43] Choi J, Ravipati A, Nimmagadda V, et al. Potential roles of PINK1 for increased PGC-1α-mediated mitochondrial fatty acid oxidation and their associations with Alzheimer's disease and diabetes, Mitochondrion. 2014; 18:41-48.

[44] Bacchetti T, Vignini A, Giulietti A, et al. Higher levels of oxidized low density lipoproteins in Alzheimer's disease patients: roles for platelet activiating factor acetyl hydrolase and paraxonase-1. Journal of Alzheimers Disease. 2015; 46:179-186.

[45] Pizzimenti S, Ciamporcero E, Daga M, et al. Interaction of aldehydes derived from lipid peroxidation and membrane proteins. Frontiers Physiology. 2013; 4:242.

[46] Reed TT. Lipid peroxidation and neurodegenerative disease. Free Radical Biology and Medicine. 2011; 51: 1302-1319.

[47] Shichiri M. The role of lipid peroxidation in neurological disorders. Journal of Clinical Biochemistry and Nutrition. 2014; 54:151-160.

[48] Magalingam KB. Radhakrishnana AK, Haleagrahran N. Protective mechanism of flavonoids in Parkinson's disease. Oxidative Medicinal Cell Longevity. 2015; 2015: 1-13.

[49] Pratico D. The neurobiology of isoprostanes and Alzheimer's disease. Biochime Bipphys Acta. 2010; 1801:930-933.

[50] Skoumalova A, Hort J. Blood markers of oxidative stress in Alzheimer's disease. Journal of Cell Molecular Medicine. 2012; 16:2291-2300.

[51] Chauhan V, Chauhan A. Oxidative stress in Alzheimer's disease. Pathophysiology. 2006; 13:195-208.

[52] Surendran S, Rajasankar S. Parkinson's disease: oxidative stress and therapeutic approaches. Neurological Science. 2010; 31:531-540

[53] Gaschler MM, Stockwell BR. Lipid peroxidation in cell death. Biochem Biophys Research Communication. 2017; 482:419-425.

[54] Takahsaki K, Nammour TM, Fukunaga M, et al. Glomerular actions of a free radical generate novel prostaglandin, 8-epi-PGF2α, in rat. Evidence for interactions with thromboxane A2 receptors. Journal of Clinical Investigation. 1992; 90:136-141.

[55] Kawikova I, Barners PJ, Takahashi T, et al. 8-epi-PGF2aplha, a novel monocylooxygenase-derived prostaglandin, constricts airways in vitro. American Journal of Respiratory Critical Care Medicine. 1996; 153:590-596.

[56] Dworski R, Murray JJ, Roberts LJI, et al. Allergen-induced synthesis of F2-isoprostanes in atopic asthmatics. American Journal of Respiratory Critical Care Medicine. 1999; 160:1947-1951.

[57] Paredi P, Kharitonov SA, Barnes PJ. Levation of echaled ethane concentration in asthma. Chest. 1997; 111:862-865.

[58] Pratico D, Basili S, Vieri M, et al. Chronic Obstructive pulmonary disease is associated with an increase in urinary levels of isoprostane G2a III, an index of oxidant stress. American journal of Respiratory critical care medicine. 1998; 158:1709-1714.

[59] Award JA, Roberts LJ. A series of prostaglandins F2-like compounds are produced in vivo in humans by a non-cyclooxygenase: use as clinical indicators of oxidant damage. Gastroenterology Clinical. 1996; 25:409-427.

[60] Roberts LJ, Morrow JD. The generation and actions of isoprostanes, Biochim Biophys Acta. 1997; 1345: 121-135.

[61] Diane LT, Tak YA, Dean PJ. The pathophysiology Significance of lipid peroxidation in oxidative cell injury. Hepatology. 1987; 7:377-387.

[62] Naik SR, Wadekar RR, Goud SS, et al. Ameliorative effects of *Tricholepisglanerrima* in experimentally induced hepatic damage in rats: Modulation of cytokines functions. Journal of Ethnopharmacology. 2015; 160: 164-172.

[63] Wadekar RR and Patil KS. Amelioration of *Trichosantheslobata* in Paracetamol-induced Hepatic Damage in Rats: A Biochemical and Histopathological Evaluation. Journal of Natural Remedies. 2017; 17: 28-37.

[64] Patil KS, Wadekar RR. Hepatoprotective activity of *Uvarianarum*in paracetamol-induced hepatic damage in rats: a biochemical and histopathological evaluation. International Journal of Pharmacognosy. 2014;1:119-129.

[65] Chao-Ling Y, Wen-Huang P, Chih-Ming H, et al. Hepatoprotective Mechanisms of Taxifolin on Carbon Tetrachloride induced Acute Liver Injury in Mice. Nutrient. 2019; 2019:1-7

Chapter 3

Thiols: Role in Oxidative Stress-Related Disorders

Abdulsamed Kükürt, Volkan Gelen, Ömer Faruk Başer, Haci Ahmet Deveci and Mahmut Karapehlivan

Abstract

The effects of oxidative stress occur as a result of peroxidative damage of the macromolecule and membranes of the cells and with the disruption of metabolic activities in the components of the cells in living organisms. Organ and tissue pathologies are known to occur when oxidative stress is excessive in the body. It is known that thiols are one of the main protective mechanisms of the body against oxidative stress. Thiols have been shown to play important roles in enzymatic reactions, apoptosis, detoxification and antioxidant protection in the body. Many studies have shown changes in thiol status and thiol/disulphide homeostasis in various diseases such as digestive system, respiratory system, reproductive system, urinary system, metabolic diseases and cancer. This also shows that the thiol state is very important in the pathogenesis of oxidative stress-mediated diseases. Therefore, it is thought that interventions that can improve thiol status may contribute to the prevention or treatment of oxidative stress-related diseases.

Keywords: antioxidant, glutathione, oxidative stress, oxidative stress-mediated disorders, taurine, thiol-disulphide homeostasis, thioredoxin

1. Introduction

Oxygen is a potentially toxic molecule, although the aerobic organisms must survive. During biochemical reactions vital to living organisms, oxygen reduced, resulting in intermediate metabolic products known as reactive oxygen species (ROS), which cause oxidative damage to many tissues. ROS is called "oxidant" or "free radical" due to the oxidative destruction they create and form in all living organisms that metabolize molecular oxygen [1]. Free radicals are very short-lived reagents, separating other electrons around high-energy electrons and disrupting their structure. Therefore, free radicals are dangerous to the organism [2, 3].

There are many defense mechanisms in the organism to prevent ROS formation and the damage caused by them. These mechanisms are generally called "antioxidant defence systems" or "antioxidants" for shortly [4]. Antioxidants serve in the body by controlling the metabolization and levels of free radicals formed as a result of normal metabolism or pathological conditions and preventing or repairing the damage that may occur by these radicals [5, 6]. In the living organism, there is a balance between the rate of formation and elimination of free radicals. This balance is called the "oxidative balance" that prevents the body from being affected by free radicals. If the oxidative balance is disturbed in favor of free radicals, oxidative stress occurs, which is one of the factors that ultimately causes damage to cells and tissues [7, 8].

All biomolecules are subject to free radical attack. But among them, lipids are the most easily affected [9]. The membranes and cell organelles that surround the cells contain a large amount of unsaturated fatty acids (PUFA). Due to the high affinity of the oxygen molecule to PUFA in the cell membrane, there is a close relationship between the two. Oxygen binds to double ligaments in PUFA found in tissues, causing lipid peroxidation [10, 11]. Lipid peroxidation is a harmful chain reaction. It can directly damage the membrane structure or damages by producing reactive aldehydes. These compounds are either metabolized at the cell or diffuse damage from initial domains to other parts. Thus, the structure of lipids in the cell membrane is disturbed, permeability for ions increases and cell death occurs [12].

Reactive nitrogen species (RNS) are another reactive species group that is as important as ROS. Nitric oxide (NO), a free radical, is the most substantial member of this group. It has the ability to directly or indirectly affect cells and tissues. As it can directly affect itself, while indirect are mediated RNS produced the interaction of NO with superoxide radicals ($O_2^{-\bullet}$) or oxygen (O_2). Most of its direct physiological effects are cyclic guanosine 3′,5’monophosphate-mediated (cGMP). It can also interact with proteins containing iron and zinc or create S-nitrosothiols through nitrosylation [2, 13–17]. Many antioxidants work in the organism to prevent damage caused by ROS and RNS. Antioxidants, present in considerably lower concentrations than the substrate, are substances that can protect an oxidation-sensitive substrate from peroxidative damage. Biological antioxidants contain all compounds that protect cellular lipids, proteins and nucleic acids from peroxidative damage. One of these compounds is thiols. Thiols play a crucial biologic role among these compounds due to their capacity to react with free radicals and their strong reducing capabilities [18].

Thiols are a member of the class of organic compounds containing sulfhydryl group (-SH). They consist of a hydrogen atom and a sulfur atom attached to a carbon atom [19]. In the organism, in the oxidation created by ROS, excess electrons pass to thiols and disulphide bonds are formed. Due to the oxidative balance, electrons in these reversible bonds can return to thiols. The antioxidant ability of thiol-disulphide homeostasis is important in enzymatic reactions, signal transduction, detoxification, transcription, regulation of enzymatic activation, cellular signaling mechanisms and apoptosis reaction [20–22]. With these in mind, in this chapter, reactive oxygen species, nitric oxide, lipid peroxidation, oxidative stress and the role of thiols in antioxidant defense is summarized and has been explained how thiol status changes in conditions associated with oxidative stress.

2. Biochemistry of reactive oxygen and nitrogen species

Free radicals and non-radical intermediates are commonly referred to as ROS. Species that contain unpaired electrons are free radicals. Species with unpaired electrons in their structure are free radicals, and because of this unpaired electron shell, free radicals have high reactivity. The most important sources of free radicals in biological systems are oxygen and nitrogen [23]. In the electron transfer chain, cells constantly convert small amounts of O_2 to ROS. ROS can be produced in many ways in the organism, including the respiratory burst that occurs in active phagocytes [24]. Respiratory burst, also known as “oxidative burst”, is the event of a rapid release of ROS such as $O_2^{-\bullet}$ and hydrogen peroxide (H_2O_2) from different cell types. Generally, these chemicals are produced by immune system cells such as neutrophils and macrophages as a result of infection of the organism by bacteria and fungi [25, 26]. In phagocytes, the respiratory burst that occurs to break down bacteria plays an important role in the immune system. $O_2^{-\bullet}$ is produced by nicotinamide adenine dinucleotide phosphate (NADPH) oxidase, a family of enzymes commonly found in many cells. In neutrophils and monocytes, myeloperoxidase is involved in combining H_2O_2 with CI-to produce hypochlorite, which plays a role in destroying bacteria [25].

Reactive oxygen species formation, as natural result of aerobic metabolism, has an important role in maintaining tissue oxygen homeostasis. $O_2^{-\bullet}$, H_2O_2 and hydroxyl ($^{\bullet}OH$) radicals are produced in mitochondria as normal metabolic by-products. Other important intracellular sources of ROS are peroxisomal enzymes, flavoprotein oxidases and microsomal cytochrome P450 enzymes [27]. ROS also play an important role in various physiological processes such as the functioning of normal vascular cells and maintenance of vessel diameter regulation [28]. It is stated that in biological systems, ROS participate in differentiation, proliferation, growth, apoptosis, cytoskeleton, migration and contraction regulation and play a role in the control of inflammatory response by stimulation of growth factor [29, 30].

Mitochondria are the main source of the $O_2^{-\bullet}$ anion most commonly found under physiological conditions [31]. The $O_2^{-\bullet}$ anion is formed by adding an electron to dioxygen. However, it is unstable because it can react spontaneously in aqueous solutions and convert into H_2O_2 and O_2 [32]. In the respiratory chain, in particular, the $O_2^{-\bullet}$ anion is formed by the leakage of electrons from complex I and III into O_2. The rate of formation depends on the number of electrons and increases with hyperoxia and high glucose concentration. The decrease in oxygen availability, acting as the final electron acceptor for complex IV, causes the accumulation of electrons. Because the $O_2^{-\bullet}$ anion is charged, it cannot pass through the membrane and remains in the mitochondrial matrix [23]. $O_2^{-\bullet}$ anion can convert to O_2 by reducing Fe^{3+} ion to Fe^{2+}. $O_2^{-\bullet}$ is detoxified with superoxide dismutase (SOD) enzymes and converted into H_2O_2 [32, 33].

Hydrogen peroxide is not a free radical, but it is mentioned in ROS because it is closely related to the detoxification or generation of free radicals [32]. It is not polar, so it can easily pass through the membranes of cells and organelles and therefore acts as a secondary messenger in a wide range of signal transduction pathways. It is detoxified into the water by catalase (CAT) and glutathione peroxidase (GPx). Imbalances in $O_2^{-\bullet}$ and H_2O_2 levels can result in the formation of $^{\bullet}OH$ radicals, which are far more dangerous than them [4]. The main source of the $^{\bullet}OH$ radical is metal-catalyzed Haber-weiss reaction [34] and the second source is the Fenton-type reaction [35]. It has been reported that the $^{\bullet}OH$ radical can react with any biological molecule in its immediate vicinity and there is no known scavenger because it is very reactive [23].

2.1 Nitric oxide

Nitric oxide is produced during the reaction which arginine is converted to citrulline catalyzed by nitric oxide synthase (NOS) which is NADPH-dependent enzyme [36, 37]. There are three isoforms of NOS: neuronal (nNOS) endothelial (eNOS) and inducible (iNOS) and it is known to be present in every cell component [17, 38]. NO is an uncharged lipophilic molecule containing unpaired electron. Although NO is not a highly reactive radical, it is important in that it can form other reactive intermediates that have an impact on protein function and the function of all organisms, as well as trigger nitrosative damage in biomolecules [39]. Therefore; it can function as an antioxidant or as an oxidant. NO, blood pressure regulator and a neurotransmitter, can produce powerful oxidants during pathological conditions [28].

The interaction of excessive amounts of $O_2^{-\bullet}$ anion with NO leads to the formation of the peroxynitrite anion ($ONOO^-$). $ONOO^-$, a cytotoxic radical, causes tissue damage and oxidizes low-density lipoproteins (LDL) [4, 40, 41]. It can also directly cause protein oxidation and DNA oxidation. $ONOO^-$ can form prooxidant nitrogen dioxide (NO_2) and $^\bullet OH$ by self-decomposition [42]. It is suggested that NO can increase the production of reactive oxygen and nitrogen species and inhibit cytochrome C oxidase in mitochondria, which can alter the activity of various processes such as respiration, mitochondrial biogenesis and oxidative stress [37]. It plays a critical role in inflammation-related carcinogenesis by activating the redox-sensitive transcription factor with nitrosative stress caused by NO, which has an important regulatory role for cellular functions. It is stated that by increasing the level of NO in plasma, it can reduce the concentration of uric acid and ascorbic acid and cause lipid peroxidation [28].

3. Lipid peroxidation

Reactive oxygen species, produced in mitochondria and extramitochondrial regions, react with polyunsaturated fatty acids (PUFA) found in complex lipids and lipoproteins, such as phospholipids found in cellular membranes, which are highly sensitive to oxidative changes. The process that causes degradation of PUFAs by chemically modified by ROS is called lipid peroxidation [43].

Lipid peroxidation in membranes is initiated by the contribution of ROS or separation of the hydrogen atom by ROS from the methylene group located between two double bonds in the PUFA. Conjugated dienes made up of PUFA react with oxygen present in the membranes at a very high rate and form a peroxyl radical ($ROO^\bullet$). Since $ROO^\bullet$ radicals are particularly highly reactive to neighboring PUFA chains, they spread the lipid peroxidation process by removing hydrogen from them. In this reaction, carbon centred radicals and lipid hydroperoxide are formed. Lipid peroxides can react with transition metal ions (iron, copper ions) to form alkoxyl radicals ($RO^\bullet$) [4, 44]. Metal ions can cause the lipid peroxide molecule to become unstable, leading to its degradation into smaller products. These products range from simple hydrocarbons to various ketones and aldehydes. The decomposition products of lipid peroxides are aldehydes such as malondialdehyde (MDA), acrolein, 4-hydroxy-2-hexenal (HHE) and 4-hydroxy-2-nonenal (HNE) [43, 45]. Commonly used lipid peroxidation markers are MDA and HNE. HNE is formed as a peroxidation product of omega-6 unsaturated fatty acids, while MDA is essentially a PUFA peroxidation product with more than two double bonds such as arachidonic acid [4].

Biomolecules undergo a lipoxidation reaction by lipid peroxidation end products such as MDA, 4-HNE and acrolein. Irreversible nonenzymatic modifications occur when these products react with lysyl (ε-NH_2), histidyl (imidazole) and cysteinyl (-SH) groups in the polypeptide chain. MDA-lysine and HNE-protein compounds formed by lipoxidation are called advanced lipoxidation end-products (ALE) [46–48].

4. Antioxidant defense system

Antioxidants, when present in low concentrations, are generally defined as substances that significantly inhibit or delay oxidative processes while they oxidize themselves, in relation to oxidizable substrates [49]. They neutralize free radicals and oxidize themselves by accepting highly reactive unpaired electrons [4]. Various transcription factors in the human body are activated or inhibited depending on the relative oxidant/antioxidant ratio. Thus, many signal paths are controlled by redox balance. The endogenous defense system consists of antioxidant compounds and specific enzymes that catalyze their antioxidant activities. There are a wide variety of powerful antioxidants that cells use, such as vitamins (C, E, A) and enzymes (CAT, GPx, SOD and thioredoxin reductase). Other non-enzymatic antioxidants available to cells include GSH, α-lipoic acid, taurine and coenzyme Q10, carotenoids and polyphenols. Especially GSH and taurine, which are thiol antioxidants, are of great importance in maintaining the redox balance [50–53].

4.1 Thiols and some of the important thiol antioxidants

Thiols are biological mercaptans (R-SH), while biological mercaptans are called biothiols. Biothiols can be classified as low molecular weight free thiols and large molecular weight protein thiols. Thiols found in biological systems play a role in the coordination of antioxidant defense systems [54]. It contains protein thiols in plasma, protein sulfhydryl groups and protein mix disulphides consisting of cysteine, cysteinylglycine, homocysteine and GSH. These thiols are also available in the form of low molecular mass disulphides, homocystine, cystinylglycine, cystine and GSSG [19]. While GSH/GSSG, especially in reduced form, consists of the low molecular weight sulfhydryl/disulphide pool inside the cell, cysteine/cystine in the form of disulphide in plasma and outside the cell as a whole [55]. It has been reported that dynamic thiol disulphide balance plays a crucial role in antioxidant system [20]. Total thiol (TT), especially protein thiol (-SH) groups in the body are considered as the main plasma antioxidants of the living organism. Most of these thiol (-SH) groups are found in albumin and constitute the major reducing groups found in body fluids [56].

Thiols play important physiological roles in processes requiring sulfur and are highly reactive, the -SH groups are readily oxidized or reduced in the presence of a catalyst [57–59]. Thiols can act as electron acceptors, reducing unstable free radicals by oxidizing, so they are powerful antioxidants. Despite their high reactivity, thiols' antioxidant potential depends on environmental, structural and catalytic factors [60–62].

Cysteine can be synthesized endogenously from methionine. Methionine, an essential amino acid in the diet, is endogenously metabolized to homocysteine and then to cysteine; Its conversion is rate limited by a few enzymes [63]. As an amino acid, cysteine has important structural roles and can bind thiol side chains to metals such as zinc, copper and iron, which are crucial for enzymatic functions. The thiol side chain of cysteine also allows it to be included in the tri-peptide thiol antioxidant GSH. Besides, cysteine metabolism through the cysteine-sulfinic acid pathway can generate taurine, although enzymatically the rate is limited, this pathway is much more complex than that of GSH [64, 65]. Both GSH and taurine are formed from cysteine with bioactive thiol groups. Although intermediate levels of cysteine are important for cellular signaling pathways, high plasma levels have been associated with cardiovascular and neurological diseases [66–68]. Additionally, high intracellular levels can increase oxidative DNA damage through the Fenton reaction [69].

4.1.1 Thioredoxin system

Thioredoxin (Trx) was first discovered in E.coli in 1964 [70]. Trx's are proteins that act as regulators in redox reactions and are found in all eukaryotic and prokaryotic organisms [71]. The Cys-Gly-Pro-Cys division is located in its active region [72].

Cytosolic thioredoxin-1 (Trx1) and mitochondrial thioredoxin-2 (Trx2) are part of the thioredoxin system, an essential and important antioxidant system for the maintenance of intracellular redox state, and play an important role in cellular redox balance and normal cell and tumor cell signaling [73, 74]. Trx exerts its antioxidant effects primarily by acting as an electron donor for peroxiredoxins. Trx is a small molecular weight protein that functions as an antioxidant by facilitating the reduction of other proteins containing the thiol (-SH) group via cysteine thiol-disulphide (-S-S-) exchange, and ribonucleotide reductase, an essential enzyme in the replication of deoxyribonucleic acid (DNA) for a hydrogen donor [75].

Thioredoxin reductases (TrxR) is a member of the flavoprotein family of pyridine nucleotide-disulphide oxidoreductases such as glutathione reductase (GSR), lipoamide dehydrogenase, mercury ion reductases [75, 76]. Members of this family include the active site in each monomer comprising the FAD, NADPH binding site and redox-active disulphide. It has a selenocysteine residue in its active site [73]. The disulphides in the active part of the TrxR reduce the substrate by catalyzing the electron transfer from NADPH to FAD. TrxRs reduce the thioredoxin protein containing two different cysteine amino acids (Trx1; Cys32 and Cys35, Trx2; Cys31 and Cys34) in its catalytic region. TrxRs have been reported to be associated with lipoic acid, lipid hydroperoxidase, cytotoxic and antibacterial polypeptide NK-lysin, dehydroascorbic acid, vitamin K, ascorbyl free radical, tumor suppressor protein p53 as well as Trx protein [71, 76–80]. It has been stated that mammalian thioredoxin reductase has three different isoenzymes, cytosolic TrxR1, mitochondrial TrxR2 and TrxR3, which is specific to testicles containing glutaredoxin region in the N terminal region [81].

Thioredoxin system has various roles in organisms and reflects the importance of the -SH group together with disulphide (-S-S-) in many reactions that are crucial in cell regulation [82]. It was previously thought that Trx was mainly involved in protecting against oxidative stress, scavenging ROS through its interaction with peroxiredoxin and working to control cellular redox balance. As a result of the studies, it has been shown that Trx contributes to redox-dependent cellular processes such as signal transduction, gene expression, apoptosis and cell growth [83, 84]. The reduced Trx binds apoptosis signal kinase-1 (ASK-1) and stops apoptosis [85]. Trx is released in response to oxidative stress and extracellular Trx exerts cytoprotective effects in inflammatory and oxidative conditions [86].

4.1.2 Glutathione system

Glutathione (GSH = γ-glutamylcysteinylglycine) is abundant in the human body. It is a tripeptide synthesized from three amino acids (cysteine, glycine and glutamate). It is a low molecular weight intracellular thiol compound and is mostly synthesized in the liver and is found in all cell types. As the regulator of intracellular redox homeostasis, most of it is stored in reduced form in the nucleus, endoplasmic reticulum, and mitochondria. The thiol group (-SH) of glutathione reduces the number of free radicals by binding to the un-shared electrons of free radicals formed as a result of oxidative stress. There are two forms in the organism: reduced (GSH) and oxidized (GSSG). The thiol-containing cysteine molecule in GSH, which is predominantly in the cell, allows ROS to take part in antioxidant roles by taking part in both degradation and removal [87–89].

The glutathione system acts as a leading cellular defense mechanism against oxidants. GSH is not only a direct ROS scavenger but also an antioxidant that has an important act in the regulation of intracellular redox status. The system consists of GPx, GSR and GSH. GSH retains its antioxidant ability in its reduced form. GPx catalyzes the reduction of H2O2 to water using GSH as a cosubstrate. GSSG is then reduced to GSH by GSR using NADPH. The cycle between these two states aids in free radical and toxic substance metabolism. The GSH/GSSG ratio is considered a sign of the redox state and relative oxidative stress level. The capability of organisms to regenerate GSH (through the synthesis of GSH or through reduction of GSSG) means the cell's success to withstand oxidative stress [90, 91].

The ability of GSH to act as an antioxidant is due to the thiol-containing cysteine part. GSH is located on both the first and second lines of ROS defense and requires GPx enzymes to catalyze the breakdown of H_2O_2 through the reduction of GSH to GSSG. GPx (GPx1), selenium-dependent, is found in the kidneys and mitochondria [92, 93]. Four other GSH peroxidases (GPx2-GPx5) have also been discovered, along with evidence of antioxidant properties in vivo [94]. Detoxified metabolites resulting from GPx defense are excreted from the cell via a glutathione S-conjugate transporter [87]. It has been reported that administration of a GSH enzyme inhibitor in rats reduces vitamin C levels in the kidney, liver, brain and lung [95]. It has been noted that GSH administration increases both vitamins C and E [96]. It is stated that vitamin C deficiency significantly decreases GSH levels in the blood [97], while vitamin C supplementation contributes to the formation of GSH [98].

4.1.3 Taurine

Cysteine can be metabolized to taurine, intracellular sulfonic acid, via cysteine-sulfinic acid. Taurine or 2-aminoethanesulfonic acid is abundant in the human body. Since there is not a carboxyl group in its structure, it is not an amino acid in theory, but it is usually referred to as proteins [99, 100]. As a result of this condition, it is released in the plasma of mammals and inside the cell [101]. Taurine is most often found where reduced O_2 molecules are produced and in locations where potentially toxic substances such as xenobiotics, retinoids and bile acids are found [102]. It is also found in high levels in white blood cells and platelets [103].

Although the mechanisms of taurine's antioxidant effects are not fully explained, possible mechanisms include regeneration of thiol groups, interfering with ROS activity and scavenging ROS [104]. It has been reported that Taurine suppresses superoxide production in mitochondria [105]. In general, taurine causes a significant reduction in ROS formation through its stimulatory effect on SOD, CAT and GPx enzyme activity [106–108]. Besides, taurine also contributes to the regulation of GSH concentrations [109]. It is thought that taurine has limited or no direct scavenging and reaction ability with ROS, and shows its antioxidant effect by increasing the activities of antioxidant enzymes such as GPx and SOD [110, 111]. It has been recorded that taurine indirectly increases endogenous GSH levels [112]. Studies have shown that taurine supplementation reduces lipid peroxidation and maintains GSH levels [113, 114].

Taurine can also inhibit free radical generation. Taurine's amino group is the direct scavenger of hypochlorous acid (HOCl) [105]. In the presence of myeloperoxidase, taurine reacts with the acid to form a less toxic oxidant, taurine chloramine (TauCl). Since neutrophils contain high levels of taurine, TauCl formation can continue as long as there is enough taurine [115]. TauCl not only plays a role in antioxidant systems by lowering HOCl levels but also inhibits O2 production and proinflammatory mediators in neutrophils and macrophages [115, 116].

5. Thiol status in oxidative stress-related various diseases

Thiol state and thiol-disulphide balance, which is an antioxidant defense system, may change due to oxidative stress in some diseases that may occur in various systems, organs and tissues in the organism.

5.1 Thiol status in digestive system diseases

In diseases of the digestive system, significant changes are observed in thiol state. For example, ROS formation in the liver increases due to alcohol intake. In this situation, serum protein thiol levels of alcohol drinkers decrease [117, 118]. It has also been determined that the level of thiol in the gallbladder increases in various gastrointestinal diseases [119]. A study showed that serum -SH levels of patients with helicobacter pylorus were significantly decreased [120]. Some studies have shown that native thiol (NT) and total thiol (TT) levels decrease and disulphide levels increase in celiac disease, acute pancreatitis, and inflammatory bowel disease [121–123]. In addition, the serum free thiol level has determined that non - alcoholic fatty liver disease (NAFLD) is associated with death from all causes in people with suspected NAFLD [124]. Impaired thiol-disulphide homeostasis has been reported in patients with hepatitis-B-induced chronic hepatitis and liver cirrhosis [125]. Again, in liver damage caused by pesticides, the thiol level was decreased, whereas black tea extract was found to improve thiol level in the liver tissue [126]. In experimental gastric damage induced by indomethacin, a non-steroidal anti-inflammatory drug, it was observed that ellagic acid treatment increased GSH levels and played a role in protecting against the harmful effects of indomethacin by reducing oxidative stress [127].

5.2 Thiol status in cardiovascular system diseases

Another situation in which thiol status changes is cardiovascular diseases. For example, in a study in preeclamptic patients characterized by high blood pressure, it was determined that the buffering function of SNO-albumin was impaired in patients in which the thiol of albumin acts as a scavenger for NO [128]. It was also observed that serum NT and TT levels of patients who had a heart attack decreased [129, 130]. In a study, it was determined that the level of mitochondria-specific thioredoxin increased, which increases NO bioavailability and reduces oxidative stress, thus protecting vascular endothelial cell function and preventing the development of atherosclerosis [131]. In rabbits, after experimental ischemia–reperfusion, it has been reported that thiol redox balance is impaired in myocardial cells and this causes abnormalities in cell function [132]. It has been reported that in case of cardiac damage caused by cyclophosphamide, thiol level decreases, but lupeol and its esters increase thiol level [133]. In sheep babesiosis, a tick-borne hemiparasitic disease, the parasite settles in the erythrocytes and causes a decrease in GSH levels in the blood. Therefore, the decrease in GSH levels indicates that excessive amounts of ROS are formed in cells [134].

5.3 Thiol status in nervous system diseases

In Parkinson's disease, oxidative stress plays an important role in the degeneration of dopaminergic neurons in the substantia nigra (SN) of patients. It was determined that the thiol antioxidant glutathione (GSH) significantly decreased in the neurons present in Substantia nigra and mitochondrial damage occurred as a result of this decrease [135, 136]. It has been observed that plasma GSH, C-SH and CG-SH levels decrease in patients with schizophrenia. However, it has been observed that Curcumin administration caused a significant increase in GSH level [137, 138]. It has been determined that TT and NT concentrations are decreased in Alzheimer's patients [139]. In the experimental Parkinson model with 6-hydroxydopamine, it was observed that the thiol level in the brain tissue decreased and the application of biarum carduchrum extract increased the thiol level [140]. In another study, hesperidin administration in 6-hydroxydopamine-induced Parkinson's model was reported to improve thiol level in brain tissue [141].

5.4 Thiol status in urinary system diseases

Studies have shown that thiol status changes in excretory system diseases. A decrease in thiol status has been reported in chronic kidney disease [142, 143]. There was a negative correlation between serum creatine level and protein thiol level. This is an indicator that serum protein thiol level will decrease in case of renal failure [144]. It has been reported that plasma protein thiol level decreases in nephrotic syndrome [145]. In another study, it was revealed that the thiol-sulphide balance decreased and this balance shifted towards disulphide in patients with acute renal failure, and the decrease in total and native thiol concentrations was associated with the severity of the disease [146]. In renal damage induced by dimethylnitrosamine, thiol level in kidney tissue decreased, whereas Simvastatin (SMN) administration improved thiol level in kidney tissue, while Thymoquinone administration was found to have no effect on thiol level [147].

5.5 Thiol status in reproductive system diseases

In polycystic ovary syndrome study, it has been observed that native thiol, total thiol, disulphide levels in the ovary tissues of patients with polycystic ovary syndrome do not change compared to the control group [148]. It has been determined that arsenic and imidocarb reduce the total thiol level in the testicular tissue of rats with testicular damage [149]. In a study, it was concluded that chemotherapeutic agents cause ovarian damage in women and that the reduction of thiol level is very important in the mechanism of this damage [150].

5.6 Thiol status in metabolic diseases

In gestational diabetes, it was determined that pregnant women with gestational diabetes have higher disulphide/natural thiol and disulphide/total thiol levels compared to healthy pregnant women [151]. In addition, in a study, in the case of diabetic nephropathy, natural thiol and total thiol levels decreased [152]. In the pathogenesis of diabetic ketoacidosis, thiol/disulphide balance changed in favor of thiol and significant decreases in disulphide level were observed [153]. Diabetic cats have been reported to have lower erythrocyte membrane thiol level than control [154]. It has been determined that thiol/disulphide homeostasis is impaired in obesity [155].

5.7 Thiol status in respiratory system diseases

In experimental asthma disease, it was determined that inflammation in the lung tissue of rats with experimental asthma increased and thiol level decreased, On the other hand, it was determined that the application of Hydro-Ethanolic Extract of *Portulaca oleracea* increased thiol level [156]. It has been reported that oxidative stress occurs during acute pulmonary inflammation induced experimentally in rats and is associated with systemic thiol homeostasis [157].

5.8 Thiol status in cancer

A study in Norway shows that thiols play a preventive role against the development of the most common breast, lung, colorectal and prostate cancers [158].

It has been determined that thiol/disulphide homeostasis plays a crucial role in the pathogenesis of cervical cancer [159]. In one study, it was reported that disruption of thiol disulphide balance is likely to contribute to the etiopathogenesis of endometrial cancer [160]. In addition, it has been stated that irregularities in thiol/disulphide homeostasis may act a part in the pathogenesis of gastric cancer, and a higher oxidative stress level may cause advanced disease to become widespread and aggressive [161].

6. Conclusion

As a result, oxidative stress can cause serious damage to the cell. Thiol is a very important antioxidant in preventing oxidative stress-induced damage and protects the cell against oxidative stress. Glutathione and taurine are among the important thiols. It is observed that thiol status changes in various diseases and thiol/disulphide homeostasis is very important in the pathogenesis of various diseases such as digestive system, respiratory system, reproductive system, urinary system, metabolic diseases and cancer. This also shows that thiol state is very important in the pathogenesis of oxidative stress-mediated diseases. Therefore, it is thought that interventions that can improve thiol status may contribute to the prevention or treatment of oxidative stress-related diseases.

Author details

Abdulsamed Kükürt[1*], Volkan Gelen[2], Ömer Faruk Başer[3], Haci Ahmet Deveci[4] and Mahmut Karapehlivan[3]

1 Department of Biochemistry, Faculty of Veterinary Medicine, Kafkas University, Kars, Turkey

2 Department of Physiology, Faculty of Veterinary Medicine, Kafkas University, Kars, Turkey

3 Department of Biochemistry, Faculty of Medicine, Kafkas University, Kars, Turkey

4 Department of Nutrition and Dietetics, Faculty of Health Sciences, Gaziantep University, Gaziantep, Turkey

*Address all correspondence to: samedkukurt@gmail.com

References

[1] Porter NA. Chemistry of lipid peroxidation. Methods Enzymol., 1984, p. 273-82. https://doi.org/10.1016/S0076-6879(84)05035-7.

[2] Ozcan A, Ogun M. Biochemistry of Reactive Oxygen and Nitrogen Species. Basic Princ. Clin. Significance Oxidative Stress, InTech; 2015. https://doi.org/10.5772/61193.

[3] Kara A, Gedikli S, Sengul E, Gelen V, Ozkanlar S. Oxidative Stress and Autophagy. Free Radicals Dis., 2016. https://doi.org/10.5772/64569.

[4] Halliwell B, Gutteridge JMC. Free Radicals in Biology and Medicine. 2015. https://doi.org/10.1093/acprof:oso/9780198717478.001.0001.

[5] Young IS. Antioxidants in health and disease. J Clin Pathol 2001;54:176-186. https://doi.org/10.1136/jcp.54.3.176.

[6] Kükürt A. Doğal bir antioksidan olarak propolis tedavisinin koruyucu etkileri. In: Evereklioğlu C, editor. Sağlık Bilim. Teor. ve Araştırmalar II, Gece Kitaplığı; 2020, p. 501-15.

[7] Birben E, Sahiner UM, Sackesen C, Erzurum S, Kalayci O. Oxidative Stress and Antioxidant Defense. World Allergy Organ J 2012;5:9-19. https://doi.org/10.1097/WOX.0b013e3182439613.

[8] Czerska M, Mikołajewska K, ZielińskiM, GromadzińskaJ, WąsowiczW. Today's oxidative stress markers. Med Pr 2015;66:393-405. https://doi.org/10.13075/mp.5893.00137.

[9] Cheeseman KH, Slater TF. An introduction to free radical biochemistry. Br Med Bull 1993;49:481-493. https://doi.org/10.1093/oxfordjournals.bmb.a072625.

[10] AyalaA, MuñozMF, ArgüellesS. Lipid Peroxidation: Production, Metabolism, and Signaling Mechanisms of Malondialdehyde and 4-Hydroxy-2-Nonenal. Oxid Med Cell Longev 2014;2014:1-31. https://doi.org/10.1155/2014/360438.

[11] Gaschler MM, Stockwell BR. Lipid peroxidation in cell death. Biochem Biophys Res Commun 2017;482:419-425. https://doi.org/10.1016/j.bbrc.2016.10.086.

[12] Romero FJ, Bosch-Morell F, RomeroMJ, JareñoEJ, RomeroB, MarínN, et al. Lipid peroxidation products and antioxidants in human disease. Environ Health Perspect 1998;106:1229-1234. https://doi.org/10.1289/ehp.98106s51229.

[13] Davis KL, Martin E, Turko I V, Murad F. Novel effects of nitric oxide. Annu Rev Pharmacol Toxicol 2001;41:203-36.:203-36.

[14] Murad F. Discovery of some of the biological effects of nitric oxide and its role in cell signaling. Biosci. Rep., vol. 24, 2004, p. 452-474. https://doi.org/10.1007/s10540-005-2741-8.

[15] Hogg N. The Biochemistry and Physiology of S-nitrosothiols. Annu Rev Pharmacol Toxicol 2002;42:585-600. https://doi.org/10.1146/annurev.pharmtox.42.092501.104328.

[16] Kükürt A, Kuru M, Karapehlivan M. Nitrik Oksit, Nitrik Oksit Sentaz ve Dişi Üreme Sistemindeki Rolleri. In: Evereklioğlu C, editor. Sağlık Bilim. Alanında Akad. Çalışmalar - II, Gece Kitaplığı; 2020, p. 113-23.

[17] Atakisi E, Merhan O. Nitric Oxide Synthase and Nitric Oxide Involvement in Different Toxicities. Nitric Oxide Synthase - Simple Enzym. Roles, InTech; 2017. https://doi.org/10.5772/intechopen.68494.

[18] Włodek L. Beneficial and harmful effects of thiols. Pol J Pharmacol 2002;54:215-223.

[19] Rossi R, Giustarini D, Milzani A, Dalle-Donne I. Cysteinylation and homocysteinylation of plasma protein thiols during ageing of healthy human beings. J Cell Mol Med 2009;13:3131-3140. https://doi.org/10.1111/j.1582-4934.2008.00417.x.

[20] Erel O, Neselioglu S. A novel and automated assay for thiol/disulphide homeostasis. Clin Biochem 2014;47:326-332. https://doi.org/10.1016/j.clinbiochem.2014.09.026.

[21] Turell L, Radi R, Alvarez B. The thiol pool in human plasma: The central contribution of albumin to redox processes. Free Radic Biol Med 2013;65:244-253. https://doi.org/10.1016/j.freeradbiomed.2013.05.050.

[22] Cremers CM, Jakob U. Oxidant Sensing by Reversible Disulfide Bond Formation. J Biol Chem 2013;288:26489-26496. https://doi.org/10.1074/jbc.R113.462929.

[23] Burton GJ, Jauniaux E. Oxidative stress. Best Pract Res Clin Obstet Gynaecol 2011;25:287-299. https://doi.org/10.1016/j.bpobgyn.2010.10.016.

[24] Alfadda AA, Sallam RM. Reactive Oxygen Species in Health and Disease. J Biomed Biotechnol 2012;2012:1-14. https://doi.org/10.1155/2012/936486.

[25] Yang P, Huang S, Xu A. The Oxidative Burst System in Amphioxus. Amphioxus Immun., Elsevier; 2016, p. 153-65. https://doi.org/10.1016/B978-0-12-849903-0.00008-7.

[26] Çenesiz S. The Role of Oxidant and Antioxidant Parameters in the Infectious Diseases: A Systematic Literature Review. Kafkas Univ Vet Fak Derg 2020;26:849-858. https://doi.org/10.9775/kvfd.2020.24618.

[27] Seifried HE, Anderson DE, Fisher EI, Milner JA. A review of the interaction among dietary antioxidants and reactive oxygen species. J Nutr Biochem 2007;18:567-579. https://doi.org/10.1016/j.jnutbio.2006.10.007.

[28] Salman KA, Ashraf S. Reactive Oxygen Species: A link between chronic inflammation and cancer. AsPac J Mol Biol Biotechnol React Oxyg Species Cancer AsPac J Mol Biol Biotechnol 2013;21:42-49.

[29] Brieger K, Schiavone S, Miller J, Krause K. Reactive oxygen species: from health to disease. Swiss Med Wkly 2012;142. https://doi.org/10.4414/smw.2012.13659.

[30] Kuru M, Kükürt A, Oral H, Öğün M. Clinical Use of Progesterone and Its Relation to Oxidative Stress in Ruminants. In: Drevenšek G, editor. Sex Horm. Neurodegener. Process. Dis., InTech; 2018, p. 303-27. https://doi.org/10.5772/intechopen.73311.

[31] Cadenas E, Davies KJA. Mitochondrial free radical generation, oxidative stress, and aging11This article is dedicated to the memory of our dear friend, colleague, and mentor Lars Ernster (1920-1998), in gratitude for all he gave to us. Free Radic Biol Med 2000;29:222-230. https://doi.org/10.1016/S0891-5849(00)00317-8.

[32] Lu J, Wang Z, Cao J, Chen Y, Dong Y. A novel and compact review on the role of oxidative stress in female reproduction. Reprod Biol Endocrinol 2018;16:80. https://doi.org/10.1186/s12958-018-0391-5.

[33] Tu BP, Weissman JS. Oxidative protein folding in eukaryotes. J Cell Biol 2004;164:341-346. https://doi.org/10.1083/jcb.200311055.

[34] Kehrer JP. The Haber–Weiss reaction and mechanisms of toxicity. Toxicology 2000;149:43-50. https://doi.org/10.1016/S0300-483X(00)00231-6.

[35] Liochev SI. The Mechanism of "Fenton-Like" Reactions and Their Importance for Biological Systems. A Biologist's View. Met. Ions Biol. Syst., Routledge; 2018, p. 1-39. https://doi.org/10.1201/9780203747605-1.

[36] Moncada S, Palmer RMJ, Higgs EA. Biosynthesis of nitric oxide from l-arginine. A pathway for the regulation of cell function and communication. Biochem Pharmacol 1989;38:1709-1715. https://doi.org/10.1016/0006-2952(89)90403-6.

[37] Bolisetty S, Jaimes E. Mitochondria and Reactive Oxygen Species: Physiology and Pathophysiology. Int J Mol Sci 2013;14:6306-6344. https://doi.org/10.3390/ijms14036306.

[38] Förstermann U, Sessa WC. Nitric oxide synthases: Regulation and function. Eur Heart J 2012;33. https://doi.org/10.1093/eurheartj/ehr304.

[39] Drew B, Leeuwenburgh C. Aging and the Role of Reactive Nitrogen Species. Ann N Y Acad Sci 2002;959:66-81. https://doi.org/10.1111/j.1749-6632.2002.tb02084.x.

[40] Kükürt A, Kuru M, Faruk Başer Ö, Karapehlivan M. Kisspeptin: Role in Female Infertility. Sex Horm. [Working Title], IntechOpen; 2020. https://doi.org/10.5772/intechopen.94925.

[41] Başer ÖF, Kükürt A, Karapehlivan M. Oksidatif stresin azaltılmasında anjiyotensin dönüştürücü enzimin rolü. In: Evereklioğlu C, editor. Sağlık Bilim. Teor. ve Araştırmalar II, Gece Kitaplığı; 2020, p. 243-53.

[42] Pacher P, Beckman JS, Liaudet L. Nitric Oxide and Peroxynitrite in Health and Disease. Physiol Rev 2007;87:315-424. https://doi.org/10.1152/physrev.00029.2006.

[43] Van Kuijk FJGM, Holte LL, Dratz EA. 4-Hydroxyhexenal: A lipid peroxidation product derived from oxidized docosahexaenoic acid. Biochim Biophys Acta - Lipids Lipid Metab 1990;1043:116-118. https://doi.org/10.1016/0005-2760(90)90118-H.

[44] Reed TT. Lipid peroxidation and neurodegenerative disease. Free Radic Biol Med 2011;51:1302-1319. https://doi.org/10.1016/j.freeradbiomed.2011.06.027.

[45] Esterbauer H, Schaur RJ, Zollner H. Chemistry and biochemistry of 4-hydroxynonenal, malonaldehyde and related aldehydes. Free Radic Biol Med 1991;11:81-128. https://doi.org/10.1016/0891-5849(91)90192-6.

[46] Lyons TJ. Glycation, Carbonyl Stress, EAGLEs, and the Vascular Complications of Diabetes. Semin Vasc Med 2002;2:175-190. https://doi.org/10.1055/s-2002-32041.

[47] Negre-Salvayre A, Coatrieux C, Ingueneau C, Salvayre R. Advanced lipid peroxidation end products in oxidative damage to proteins. Potential role in diseases and therapeutic prospects for the inhibitors. Br J Pharmacol 2008;153:6-20. https://doi.org/10.1038/sj.bjp.0707395.

[48] Dalle-Donne I, Giustarini D, Colombo R, Rossi R, Milzani A. Protein carbonylation in human diseases. Trends Mol Med 2003;9:169-176. https://doi.org/10.1016/S1471-4914(03)00031-5.

[49] Vaya J, Aviram M. Nutritional Antioxidants Mechanisms of Action, Analyses of Activities and Medical Applications. Curr Med Chem Endocr Metab Agents 2001;1:99-117. https://doi.org/10.2174/1568013013359168.

[50] Biswas SK, Newby DE, Rahman I, Megson IL. Depressed glutathione synthesis precedes oxidative stress and atherogenesis in Apo-E−/− mice. Biochem Biophys Res Commun 2005;338:1368-1373. https://doi.org/10.1016/j.bbrc.2005.10.098.

[51] Packer L, Witt EH, Tritschler HJ. Alpha-lipoic acid as a biological antioxidant. Free Radic Biol Med 1995;19:227-250. https://doi.org/10.1016/0891-5849(95)00017-R.

[52] Ernster L, Forsmark-Andree P. Ubiquinol: an endogenous antioxidant in aerobic organisms. Clin Investig 1993;71:60-65. https://doi.org/10.1007/BF00226842.

[53] Hanschmann E-M, Godoy JR, Berndt C, Hudemann C, Lillig CH. Thioredoxins, Glutaredoxins, and Peroxiredoxins—Molecular Mechanisms and Health Significance: from Cofactors to Antioxidants to Redox Signaling. Antioxid Redox Signal 2013;19:1539-1605. https://doi.org/10.1089/ars.2012.4599.

[54] Sen CK, Packer L. Thiol homeostasis and supplements in physical exercise. Am J Clin Nutr 2000;72:653-669. https://doi.org/10.1093/ajcn/72.2.653S.

[55] Dalle–Donne I, Milzani A, Gagliano N, Colombo R, Giustarini D, Rossi R. Molecular Mechanisms and Potential Clinical Significance of S -Glutathionylation. Antioxid Redox Signal 2008;10:445-474. https://doi.org/10.1089/ars.2007.1716.

[56] Prakash M, Shetty MS, Tilak P, Anwar N. Total Thiols : Biomedical Importance And Their Alteration In Various Disorders. Online J Heal Allied Sci 2009;8:1-9.

[57] Dhakshinamoorthy A, Alvaro M, Garcia H. Aerobic oxidation of thiols to disulfides using iron metal–organic frameworks as solid redox catalysts. Chem Commun 2010;46:6476. https://doi.org/10.1039/c0cc02210a.

[58] Giles NM, Watts AB, Giles GI, Fry FH, Littlechild JA, Jacob C. Metal and Redox Modulation of Cysteine Protein Function. Chem Biol 2003;10:677-693. https://doi.org/10.1016/S1074-5521(03)00174-1.

[59] Kachur A V., Koch CJ, Biaglow JE. Mechanism of Copper-Catalyzed Oxidation of Glutathione. Free Radic Res 1998;28:259-269. https://doi.org/10.3109/10715769809069278.

[60] Barron ESG. Thiol Groups of Biological Importance. Adv. Enzymol. Relat. Areas Mol. Biol., 2006, p. 201-66. https://doi.org/10.1002/9780470122563.ch4.

[61] Conte M Lo, Carroll KS. The Chemistry of Thiol Oxidation and Detection. Oxidative Stress Redox Regul., Dordrecht: Springer Netherlands; 2013, p. 1-42. https://doi.org/10.1007/978-94-007-5787-5_1.

[62] Winterbourn CC, Metodiewa D. Reactivity of biologically important thiol compounds with superoxide and hydrogen peroxide. Free Radic Biol Med 1999;27:322-328. https://doi.org/10.1016/S0891-5849(99)00051-9.

[63] Ganguly P, Alam SF. Role of homocysteine in the development of cardiovascular disease. Nutr J 2015;14:6. https://doi.org/10.1186/1475-2891-14-6.

[64] Bella DL, Hahn C, Stipanuk MH. Effects of nonsulfur and sulfur amino acids on the regulation of hepatic enzymes of cysteine metabolism. Am J Physiol Metab 1999;277:E144–E153. https://doi.org/10.1152/ajpendo.1999.277.1.E144.

[65] Lee J-I, Londono M, Hirschberger LL, Stipanuk MH. Regulation of cysteine dioxygenase and γ-glutamylcysteine synthetase is associated with hepatic cysteine level. J Nutr Biochem 2004;15:112-122. https://doi.org/10.1016/j.jnutbio.2003.10.005.

[66] El-Khairy L, Ueland PM, Nygård O, Refsum H, Vollset SE. Lifestyle and cardiovascular disease risk factors as

determinants of total cysteine in plasma: the Hordaland Homocysteine Study. Am J Clin Nutr 1999;70:1016-1024. https://doi.org/10.1093/ajcn/70.6.1016.

[67] Özkan Y, Özkan E, Şimşek B. Plasma total homocysteine and cysteine levels as cardiovascular risk factors in coronary heart disease. Int J Cardiol 2002;82:269-277. https://doi.org/10.1016/S0167-5273(02)00010-4.

[68] Heafield MT, Fearn S, Steventon GB, Waring RH, Williams AC, Sturman SG. Plasma cysteine and sulphate levels in patients with motor neurone, Parkinson's and Alzheimer's disease. Neurosci Lett 1990;110:216-220. https://doi.org/10.1016/0304-3940(90)90814-P.

[69] Park S, Imlay JA. High Levels of Intracellular Cysteine Promote Oxidative DNA Damage by Driving the Fenton Reaction. J Bacteriol 2003;185:1942-1950. https://doi.org/10.1128/JB.185.6.1942-1950.2003.

[70] Laurent TC, Moore EC, Reichard P. Enzymatic Synthesis of Deoxyribonucleotides. J Biol Chem 1964;239:3436-3444. https://doi.org/10.1016/S0021-9258(18)97742-2.

[71] Arnér ESJ, Holmgren A. Physiological functions of thioredoxin and thioredoxin reductase. Eur J Biochem 2000;267:6102-6109. https://doi.org/10.1046/j.1432-1327.2000.01701.x.

[72] Takagi Y, Mitsui A, Nishiyama A, Nozaki K, Sono H, Gon Y, et al. Overexpression of thioredoxin in transgenic mice attenuates focal ischemic brain damage. Proc Natl Acad Sci 1999;96:4131-4136. https://doi.org/10.1073/pnas.96.7.4131.

[73] Mustacich D, Powis G. Thioredoxin reductase. Biochem J 2000;346:1. https://doi.org/10.1042/0264-6021:3460001.

[74] Monteiro HP, Ogata FT, Stern A. Thioredoxin promotes survival signaling events under nitrosative/oxidative stress associated with cancer development. Biomed J 2017;40:189-199. https://doi.org/10.1016/j.bj.2017.06.002.

[75] Holmgren A. Thioredoxin and Glutaredoxin Systems. J Biol Chem 1989;264:13963-13966. https://doi.org/10.1016/S0021-9258(18)71625-6.

[76] Nordberg J, Arnér ESJ. Reactive oxygen species, antioxidants, and the mammalian thioredoxin system. Free Radic Biol Med 2001;31:1287-1312. https://doi.org/10.1016/S0891-5849(01)00724-9.

[77] Arnér ESJ, Nordberg J, Holmgren A. Efficient Reduction of Lipoamide and Lipoic Acid by Mammalian Thioredoxin Reductase. Biochem Biophys Res Commun 1996;225:268-274. https://doi.org/10.1006/bbrc.1996.1165.

[78] Andersson M, Holmgren A, Spyrou G. NK-lysin, a Disulfide-containing Effector Peptide of T-lymphocytes, Is Reduced and Inactivated by Human Thioredoxin Reductase. J Biol Chem 1996;271:10116-10120. https://doi.org/10.1074/jbc.271.17.10116.

[79] May JM, Mendiratta S, Hill KE, Burk RF. Reduction of Dehydroascorbate to Ascorbate by the Selenoenzyme Thioredoxin Reductase. J Biol Chem 1997;272:22607-22610. https://doi.org/10.1074/jbc.272.36.22607.

[80] Casso D, Beach D. A mutation in a thioredoxin reductase homolog suppresses p53-induced growth inhibition in the fission yeast. MGG Mol Gen Genet 1996;252:518. https://doi.org/10.1007/s004380050259.

[81] Arnér ESJ. Focus on mammalian thioredoxin reductases — Important selenoproteins with versatile functions.

Biochim Biophys Acta - Gen Subj 2009;1790:495-526. https://doi.org/10.1016/j.bbagen.2009.01.014.

[82] Whayne TF, Parinandi N, Maulik N. Thioredoxins in cardiovascular disease. Can J Physiol Pharmacol 2015;93:903-911. https://doi.org/10.1139/cjpp-2015-0105.

[83] Rhee SG, Chae HZ, Kim K. Peroxiredoxins: A historical overview and speculative preview of novel mechanisms and emerging concepts in cell signaling. Free Radic Biol Med 2005;38:1543-1552. https://doi.org/10.1016/j.freeradbiomed.2005.02.026.

[84] Matsuo Y, Yodoi J. Extracellular thioredoxin: A therapeutic tool to combat inflammation. Cytokine Growth Factor Rev 2013;24:345-353. https://doi.org/10.1016/j.cytogfr.2013.01.001.

[85] Saitoh M. Mammalian thioredoxin is a direct inhibitor of apoptosis signal-regulating kinase (ASK) 1. EMBO J 1998;17:2596-2606. https://doi.org/10.1093/emboj/17.9.2596.

[86] Nakamura H, Masutani H, Yodoi J. Extracellular thioredoxin and thioredoxin-binding protein 2 in control of cancer. Semin Cancer Biol 2006;16:444-451. https://doi.org/10.1016/j.semcancer.2006.09.001.

[87] Hayes JD, McLellan LI. Glutathione and glutathione-dependent enzymes represent a co-ordinately regulated defence against oxidative stress. Free Radic Res 1999;31:273-300. https://doi.org/10.1080/10715769900300851.

[88] Maher P, Lewerenz J, Lozano C, Torres JL. A novel approach to enhancing cellular glutathione levels. J Neurochem 2008;107:690-700. https://doi.org/10.1111/j.1471-4159.2008.05620.x.

[89] Masella R, Di Benedetto R, Varì R, Filesi C, Giovannini C. Novel mechanisms of natural antioxidant compounds in biological systems: involvement of glutathione and glutathione-related enzymes. J Nutr Biochem 2005;16:577-586. https://doi.org/10.1016/j.jnutbio.2005.05.013.

[90] Curtin JF, Donovan M, Cotter TG. Regulation and measurement of oxidative stress in apoptosis. J Immunol Methods 2002;265:49-72. https://doi.org/10.1016/S0022-1759(02)00070-4.

[91] Jones DP. Redox potential of GSH/GSSG couple: Assay and biological significance. Methods Enzymol., 2002, p. 93-112. https://doi.org/10.1016/S0076-6879(02)48630-2.

[92] Rotruck JT, Pope AL, Ganther HE, Swanson AB, Hafeman DG, Hoekstra WG. Selenium: Biochemical Role as a Component of Glutathione Peroxidase. Science (80-) 1973;179:588-90. https://doi.org/10.1126/science.179.4073.588.

[93] Esworthy RS, Ho Y-S, Chu F-F. The Gpx1 Gene Encodes Mitochondrial Glutathione Peroxidase in the Mouse Liver. Arch Biochem Biophys 1997;340:59-63. https://doi.org/10.1006/abbi.1997.9901.

[94] Arai M, Imai H, Koumura T, Yoshida M, Emoto K, Umeda M, et al. Mitochondrial Phospholipid Hydroperoxide Glutathione Peroxidase Plays a Major Role in Preventing Oxidative Injury to Cells. J Biol Chem 1999;274:4924-4933. https://doi.org/10.1074/jbc.274.8.4924.

[95] Mãrtensson J, Meister A. Glutathione deficiency decreases tissue ascorbate levels in newborn rats: Ascorbate spares glutathione and protects. Proc Natl Acad Sci U S A 1991;88:4656-4660. https://doi.org/10.1073/pnas.88.11.4656.

[96] Chan AC, Tran K, Raynor T, Ganz P, Chow CK. Regeneration of

vitamin E in human platelets. Free Radic Biol Med 1990;9:11. https://doi.org/10.1016/0891-5849(90)90206-X.

[97] Henning SM, Zhang JZ, McKee RW, Swendseid ME, Jacob RA. Glutathione Blood Levels and Other Oxidant Defense Indices in Men Fed Diets Low in Vitamin C. J Nutr 1991;121:1969-1975. https://doi.org/10.1093/jn/121.12.1969.

[98] Johnston CS, Meyer CG, Srilakshmi JC. Vitamin C elevates red blood cell glutathione in healthy adults. Am J Clin Nutr 1993;58:103-105. https://doi.org/10.1093/ajcn/58.1.103.

[99] Kendler BS. Taurine: An overview of its role in preventive medicine. Prev Med (Baltim) 1989;18:79-100. https://doi.org/10.1016/0091-7435(89)90056-X.

[100] Geggel HS, Ament ME, Heckenlively JR, Martin DA, Kopple JD. Nutritional Requirement for Taurine in Patients Receiving Long-Term Parenteral Nutrition. N Engl J Med 1985;312:142-146. https://doi.org/10.1056/NEJM198501173120302.

[101] Brosnan JT, Brosnan ME. The Sulfur-Containing Amino Acids: An Overview. J Nutr 2006;136:1636S–1640S. https://doi.org/10.1093/jn/136.6.1636S.

[102] Wright CE, Tallan HH, Lin YY. Taurine: Biological Update. Annu Rev Biochem 1986;55:427-453. https://doi.org/10.1146/annurev.bi.55.070186.002235.

[103] Franconi F, Di Leo MAS, Bennardini F, Ghirlanda G. Is Taurine Beneficial in Reducing Risk Factors for Diabetes Mellitus? Neurochem Res 2004;29:143-150.https://doi.org/10.1023/B:NERE.0000010443.05899.2f.

[104] Imae M, Asano T, Murakami S. Potential role of taurine in the prevention of diabetes and metabolic syndrome. Amino Acids 2014;46:81-88. https://doi.org/10.1007/s00726-012-1434-4.

[105] Schaffer SW, Azuma J, Mozaffari M. Role of antioxidant activity of taurine in diabetes. Can J Physiol Pharmacol 2009;87:91-99. https://doi.org/10.1139/Y08-110.

[106] Wang G, Li W, Lu X, Zhao X, Xu L. Taurine attenuates oxidative stress and alleviates cardiac failure in type I diabetic rats. Croat Med J 2013;54:171-179. https://doi.org/10.3325/cmj.2013.54.171.

[107] Shivananjappa MM, Muralidhara. Taurine attenuates maternal and embryonic oxidative stress in a streptozotocin-diabetic rat model. Reprod Biomed Online 2012;24:558-566. https://doi.org/10.1016/j.rbmo.2012.01.016.

[108] Rashid K, Das J, Sil PC. Taurine ameliorate alloxan induced oxidative stress and intrinsic apoptotic pathway in the hepatic tissue of diabetic rats. Food Chem Toxicol 2013;51:317-329. https://doi.org/10.1016/j.fct.2012.10.007.

[109] Kim YG, Kim SK, Kwon JW, Park OJ, Kim SG, Kim YC, et al. Effects of cysteine on amino acid concentrations and transsulfuration enzyme activities in rat liver with protein–calorie malnutrition. Life Sci 2003;72:1171-1181. https://doi.org/10.1016/S0024-3205(02)02366-4.

[110] Aruoma OI, Halliwell B, Hoey BM, Butler J. The antioxidant action of taurine, hypotaurine and their metabolic precursors. Biochem J 1988;256:251-255. https://doi.org/10.1042/bj2560251.

[111] Vohra BPS, Hui X. Taurine Protects against Carbon Tetrachloride Toxicity in the Cultured Neurons and In Vivo. Arch Physiol Biochem 2001;109:90-94. https://doi.org/10.1076/apab.109.1.90.4287.

[112] Noeman SA, Hamooda HE, Baalash AA. Biochemical Study of Oxidative Stress Markers in the Liver, Kidney and Heart of High Fat Diet Induced Obesity in Rats. Diabetol Metab Syndr 2011;3:17. https://doi.org/10.1186/1758-5996-3-17.

[113] Anand P, Rajakumar D, Jeraud M, William Fe AJ, Balasubram T. Effects of Taurine on Glutathione Peroxidase, Glutathione Reductase and Reduced Glutathione Levels in Rats. Pakistan J Biol Sci 2011;14:219-225. https://doi.org/10.3923/pjbs.2011.219.225.

[114] Oudit GY, Trivieri MG, Khaper N, Husain T, Wilson GJ, Liu P, et al. Taurine Supplementation Reduces Oxidative Stress and Improves Cardiovascular Function in an Iron-Overload Murine Model. Circulation 2004;109:1877-1885. https://doi.org/10.1161/01.CIR.0000124229.40424.80.

[115] Schuller-Levis GB, Park E. Taurine and Its Chloramine: Modulators of Immunity. Neurochem Res 2004;29:117-126. https://doi.org/10.1023/B:NERE.0000010440.37629.17.

[116] Marcinkiewicz J, Grabowska A, Bereta J, Stelmaszynska T. Taurine chloramine, a product of activated neutrophils, inhibits in vitro the generation of nitric oxide and other macrophage inflammatory mediators. J Leukoc Biol 1995;58:667-674. https://doi.org/10.1002/jlb.58.6.667.

[117] Prakash M, K. Shetty J, Tripathy S, Verma M, Vasudev S, V. Bhandar P. Serum Total Thiol Status in Alcohol Abusers. Asian J Biochem 2007;3:48-51. https://doi.org/10.3923/ajb.2008.48.51.

[118] Venkatraman A, Landar A, Davis AJ, Ulasova E, Page G, Murphy MP, et al. Oxidative modification of hepatic mitochondria protein thiols: effect of chronic alcohol consumption. Am J Physiol Liver Physiol 2004;286:G521–G527. https://doi.org/10.1152/ajpgi.00399.2003.

[119] Yılmaz İA, Akçay T, Çakatay U, Telci A, Ataus S, Yalçın V. Relation between bladder cancer and protein oxidation. Int Urol Nephrol 2003;35:345-350. https://doi.org/10.1023/B:UROL.0000022920.93994.ba.

[120] Aslan M, Nazligul Y, Horoz M, Bolukbas C, Bolukbas FF, Gur M, et al. Serum paraoxonase-1 activity in Helicobacter pylori infected subjects. Atherosclerosis 2008;196:270-274. https://doi.org/10.1016/j.atherosclerosis.2006.10.024.

[121] Kaplan M, Ates I, Yuksel M, Ozin YO, Alisik M, Erel O, et al. Thiol/disulphide homeostasis in celiac disease. World J Gastrointest Pharmacol Ther 2017;8:120. https://doi.org/10.4292/wjgpt.v8.i2.120.

[122] Yuksel M, Ates I, Kaplan M, Alışık M, Erel Ö, Saygılı F, et al. The dynamic thiol/disulphide homeostasis in inflammatory bowel disease and its relation with disease activity and pathogenesis. Int J Colorectal Dis 2016;31:1229-1231. https://doi.org/10.1007/s00384-015-2439-8.

[123] Koseoglu H, Alisik M, Basaran M, Tayfur Yurekli O, Solakoglu T, Tahtaci M, et al. Dynamic thiol/disulphide homeostasis in acute pancreatitis. Turkish J Gastroenterol 2018:348-453. https://doi.org/10.5152/tjg.2018.17499.

[124] Damba T, Bourgonje AR, Abdulle AE, Pasch A, Sydor S, Berg EH, et al. Oxidative stress is associated with suspected non-alcoholic fatty liver disease and all-cause mortality in the general population. Liver Int 2020;40:2148-2159. https://doi.org/10.1111/liv.14562.

[125] Dertli R, Keskin M, Bıyık M, Ataseven H, Polat H, Demir A, et al.

Dynamic thiol-disulfide homeostasis is disturbed in hepatitis B virus-related chronic hepatitis and liver cirrhosis. TURKISH J Med Sci 2018;48:985-992. https://doi.org/10.3906/sag-1803-135.

[126] Khan SM. Protective effect of black tea extract on the levels of lipid peroxidation and antioxidant enzymes in liver of mice with pesticide-induced liver injury. Cell Biochem Funct 2006;24:327-332. https://doi.org/10.1002/cbf.1246.

[127] Kaya İ, Kaya MM, Kükürt A, Özcan A, Karaman M, Deveci HA, et al. Effect of Ellagic Acid on Some Oxidative Stress Parameters and Cyclooxygenase-2 Reactivity in Mice with Experimental Gastric Injury. Japanese J Gastroenterol Hepatol 2019;2:1-9.

[128] Gandley RE, Tyurin VA, Huang W, Arroyo A, Daftary A, Harger G, et al. S -Nitrosoalbumin–Mediated Relaxation Is Enhanced by Ascorbate and Copper. Hypertension 2005;45:21-27. https://doi.org/10.1161/01.HYP.0000150158.42620.3e.

[129] Altıparmak IH, Erkuş ME, Sezen H, Demirbag R, Gunebakmaz O, Kaya Z, et al. The relation of serum thiol levels and thiol/disulphide homeostasis with the severity of coronary artery disease. Kardiol Pol 2016:1346-1353. https://doi.org/10.5603/KP.a2016.0085.

[130] Kundi H, Ates I, Kiziltunc E, Cetin M, Cicekcioglu H, Neselioglu S, et al. A novel oxidative stress marker in acute myocardial infarction; thiol/disulphide homeostasis. Am J Emerg Med 2015;33:1567-1571. https://doi.org/10.1016/j.ajem.2015.06.016.

[131] Patwari P, Lee RT. Thioredoxins, Mitochondria, and Hypertension. Am J Pathol 2007;170:805-808. https://doi.org/10.2353/ajpath.2007.061243.

[132] Ceconi C. New insights on myocardial pyridine nucleotides and thiol redox state in ischemia and reperfusion damage. Cardiovasc Res 2000;47:586-594. https://doi.org/10.1016/S0008-6363(00)00104-8.

[133] Sudharsan PT, Mythili Y, Selvakumar E, Varalakshmi P. Lupeol and its ester ameliorate the cyclophosphamide provoked cardiac lysosomal damage studied in rat. Mol Cell Biochem 2006;282:23-29. https://doi.org/10.1007/s11010-006-1169-1.

[134] Deveci HA, Nur G, Kukurt A. Biochemical and histopathological changes of babesiosis in naturally infected sheep in gaziantep region. Fresenius Environ Bull 2017;26:4883-4889.

[135] Jagatha B, Mythri RB, Vali S, Bharath MMS. Curcumin treatment alleviates the effects of glutathione depletion in vitro and in vivo: Therapeutic implications for Parkinson's disease explained via in silico studies. Free Radic Biol Med 2008;44:907-917. https://doi.org/10.1016/j.freeradbiomed.2007.11.011.

[136] Anderson CC, Marentette JO, Rauniyar AK, Prutton KM, Khatri M, Matheson C, et al. Maneb alters central carbon metabolism and thiol redox status in a toxicant model of Parkinson's disease. Free Radic Biol Med 2021;162:65-76. https://doi.org/10.1016/j.freeradbiomed.2020.11.028.

[137] Dietrich-Muszalska A, Olas B, Głowacki R, Bald E. Oxidative/Nitrative Modifications of Plasma Proteins and Thiols from Patients with Schizophrenia. Neuropsychobiology 2009;59:1-7. https://doi.org/10.1159/000202822.

[138] Fendri C, Mechri A, Khiari G, Othman A, Kerkeni A, Gaha L. Implication du stress oxydant dans la physiopathologie de la schizophrénie: revue de la literature. Encephale

2006;32:244-252. https://doi.org/10.1016/S0013-7006(06)76151-6.

[139] Gumusyayla S, Vural G, Bektas H, Deniz O, Neselioglu S, Erel O. A novel oxidative stress marker in patients with Alzheimer's disease: dynamic thiol–disulphide homeostasis. Acta Neuropsychiatr 2016;28:315-320. https://doi.org/10.1017/neu.2016.13.

[140] Zanganehnejad Z, Setorki M. Effect of Biarum carduchrum extract on brain tissue thiol level in rat model of 6-hydroxydopamine-induced Parkinson's disease. J Herbmed Pharmacol 2018;7:136-140. https://doi.org/10.15171/jhp.2018.23.

[141] Antunes MS, Goes ATR, Boeira SP, Prigol M, Jesse CR. Protective effect of hesperidin in a model of Parkinson's disease induced by 6-hydroxydopamine in aged mice. Nutrition 2014;30:1415-1422. https://doi.org/10.1016/j.nut.2014.03.024.

[142] Galle J. Oxidative stress in chronic renal failure. Nephrol Dial Transplant 2001;16:2135-2137. https://doi.org/10.1093/ndt/16.11.2135.

[143] Himmelfarb J, Hakim RM. Oxidative stress in uremia. Curr Opin Nephrol Hypertens 2003;12:593-598. https://doi.org/10.1097/00041552-200311000-00004.

[144] Prakash M, Shetty J, Rao L, Sharma S, Rodrigues A, Prabhu R. Serum paraoxonase activity and protein thiols in chronic renal failure patients. Indian J Nephrol 2008;18:13. https://doi.org/10.4103/0971-4065.41282.

[145] Karthikeyan K, Sinha I, Prabhu K, Bhaskaranand N, Rao A. Plasma Protein Thiols and Total Antioxidant Power in Pediatric Nephrotic Syndrome. Nephron Clin Pract 2008;110:c10–c14. https://doi.org/10.1159/000148210.

[146] Otal Y, Demircan S, Sener A, Alısık M, Tanrıverdi F, Haydar FGE, et al. Acute Renal Failure and Thiol-Disulfide Homeostasis. J Nephrol Ther 2018;08. https://doi.org/10.4172/2161-0959.1000312.

[147] Saricicek E, Celik A, Uremis N, Kilinc M. Protective effects of simvastatin, Nigella sativa oil and thmoquinone against dimethylnitrosamine-induced oxidative stress in rat kidney. Biomed Res 2016;27.

[148] Aydın GA, Turan Özsoy HG, Ankaralı H, Özgen G, Neşelioğlu S. The association of dynamic thiol-disulfide homeostasis and inflammatory markers in patients with polycystic ovary syndrome. Taiwan J Obstet Gynecol 2020;59:79-84. https://doi.org/10.1016/j.tjog.2019.11.012.

[149] Mahajan L, Verma PK, Raina R, Sood S. Potentiating effect of imidacloprid on arsenic-induced testicular toxicity in Wistar rats. BMC Pharmacol Toxicol 2018;19:48. https://doi.org/10.1186/s40360-018-0239-9.

[150] Spears N, Lopes F, Stefansdottir A, Rossi V, De Felici M, Anderson RA, et al. Ovarian damage from chemotherapy and current approaches to its protection. Hum Reprod Update 2019;25:673-693. https://doi.org/10.1093/humupd/dmz027.

[151] Gürlek B, Alan M, Çolak S, Önal Ö, Erel Ö, Biçer C. Dynamic thiol/disulfide homeostasis in gestational diabetes mellitus: Is it related with adverse perinatal outcomes? Med Sci Discov 2019;6:198-204. https://doi.org/10.36472/msd.v6i9.293.

[152] Eren MA, Koyuncu İ, İncebıyık H, Karakaş H, Erel Ö, Sabuncu T. The evaluation of thiol/disulphide homeostasis in diabetic nephropathy. Diabetes Res Clin Pract 2019;148:249-253. https://doi.org/10.1016/j.diabres.2019.01.022.

[153] Otal Y, Kahraman FA, Haydar FG, Erel Ö. Dynamic thiol/disulphide

homeostasis as oxidative stress marker in diabetic ketoacidosis. TURKISH J Med Sci 2021. https://doi.org/10.3906/sag-1904-55.

[154] Zini E, Gabai G, Salesov E, Gerardi G, Da Dalt L, Lutz TA, et al. Oxidative status of erythrocytes, hyperglycemia, and hyperlipidemia in diabetic cats. J Vet Intern Med 2020;34:616-625. https://doi.org/10.1111/jvim.15732.

[155] Mengen E, Uçaktürk SA, Kocaay P, Kaymaz Ö, Neşelioğlu S, Erel Ö. The Significance of Thiol/Disulfide Homeostasis and Ischemia-modified Albumin Levels in Assessing Oxidative Stress in Obese Children and Adolescents. J Clin Res Pediatr Endocrinol 2020;12:45-54. https://doi.org/10.4274/jcrpe.galenos.2019.2019.0039.

[156] Boskabady MH, Kaveh M, Shakeri F, Mohammadian Roshan N, Rezaee R. Hydro-ethanolic extract of portulaca oleracea ameliorates total and differential wbc, lung pathology and oxidative biomarkers in asthmatic rats. Iran J Pharm Res 2019;18:1947-1958. https://doi.org/10.22037/ijpr.2019.13712.11817.

[157] Cotgreave IA, Johansson U, Moldéus P, Brattsand R. Lung and systematic thiol homeostasis during an acute lung inflammation in the rat. Toxicology 1988;50:331-343. https://doi.org/10.1016/0300-483X(88)90048-0.

[158] Gào X, Wilsgaard T, Jansen EHJM, Xuan Y, Anusruti A, Brenner H, et al. Serum total thiol levels and the risk of lung, colorectal, breast and prostate cancer: A prospective case–cohort study. Int J Cancer 2020;146:1261-1267. https://doi.org/10.1002/ijc.32428.

[159] Sezgin B, Kinci MF, Pirinççi F, Camuzcuoğlu A, Erel Ö, Neşelioğlu S, et al. <scp>Thiol-disulfide</scp> status of patients with cervical cancer. J Obstet Gynaecol Res 2020;46:2423-2429. https://doi.org/10.1111/jog.14480.

[160] Sezgin B, Pirinççi F, Camuzcuoğlu A, Erel Ö, Neşelioğlu S, Camuzcuoğlu H. Assessment of thiol disulfide balance in early-stage endometrial cancer. J Obstet Gynaecol Res 2020;46:1140-1147. https://doi.org/10.1111/jog.14301.

[161] Hizal M, Sendur MAN, Bilgin B, Akinci MB, Dede DS, Neselioglu S, et al. Evaluation of dynamic serum thiol/disulfide homeostasis in locally advanced and metastatic gastric cancer. J Oncol Sci 2018;4:1-4. https://doi.org/10.1016/j.jons.2018.01.002.

Chapter 4

Lipid Peroxidation as a Link between Unhealthy Diets and the Metabolic Syndrome

Arnold N. Onyango

Abstract

Unhealthy diets, such as those high in saturated fat and sugar accelerate the development of non-communicable diseases. The metabolic syndrome is a conglomeration of disorders such as abdominal obesity, hypertension, impaired glucose regulation and dyslipidemia, which increases the risk for diabetes and cardiovascular disease. The prevalence of the metabolic syndrome is increasing globally, and dietary interventions may help to reverse this trend. A good understanding of its pathophysiological mechanisms is needed for the proper design of such interventions. This chapter discusses how lipid peroxidation is associated with the development of this syndrome, mainly through the formation of bioactive aldehydes, such as 4-hydroxy-2-nonenal, malondialdehyde, acrolein and glyoxal, which modify biomolecules to induce cellular dysfunction, including the enhancement of oxidative stress and inflammatory signaling. It gives a current understanding of the mechanisms of formation of these aldehydes and how dietary components such as saturated fatty acids promote oxidative stress, leading to lipid oxidation. It also outlines mechanisms, apart from free radical scavenging and singlet oxygen quenching, by which various dietary constituents prevent oxidative stress and lipid oxidation in vivo.

Keywords: Oxidative stress, lipid peroxidation, insulin resistance; metabolic syndrome

1. Introduction

The metabolic syndrome (MS) refers to the occurrence in an individual of multiple physiological disorders related to obesity, hypertension, dysregulated blood glucose and dysregulated blood lipids, and is a risk factor for diabetes and cardiovascular disease [1]. It has been defined more specifically, and in slightly different ways by the National Cholesterol Education Program (NCEP) Adult Treatment Panel III, and by the World Health Organization (WHO). According to the former, MS is characterized by at least three of the following five clinical or biochemical abnormalities: abdominal obesity, arterial hypertension, elevated fasting blood glucose, high plasma triglycerides, and reduced high density lipoprotein cholesterol (HDL-c) [2]. On the other hand, WHO defined it as the occurrence of impaired glucose tolerance or impaired fasting glucose or diabetes and any two of the following: hypertension; elevated trigycerides or low HDL-c; abdominal obesity or obesity as determined by BMI; or microalbuminaria [1].

A proper understanding of the etiology of MS is necessary for its prevention and treatment. This chapter focuses on the role of lipid peroxidation in this pathophysiological process. It begins with a current understanding of the mechanisms of lipid oxidation, with emphasis on the formation of highly reactive lipid oxidation products such as 4-hydroxy-2-nonenal, malondialdehyde, acrolein and glyoxal. This is followed by a discussion of how these aldehydes and other lipid oxidation products contribute to the different MS components. The role of major dietary components in the initiation of oxidative stress and lipid oxidation, as well as mechanisms by which specific dietary components inhibit such undesirable events are also discussed.

2. Mechanisms of lipid peroxidation (LPO) and the formation of bioactive lipid oxidation products

In cells, extensive lipid oxidation and the accumulation of lipid oxidation products occurs under conditions of oxidative stress, when the concentrations of reactive oxygen species (ROS) such as superoxide anions, hydrogen peroxide and singlet oxygen increase, and are not matched by an increase in the cellular antioxidant capacity [3]. Electron leakage from the mitochondrial electron transport chain, or the actions of enzymes such as NADPH oxidases and xanthine oxidase generate superoxide anions ($^{\cdot-}O_2$), which are converted by superoxide dismutase to hydrogen peroxide (H_2O_2), which may be converted by ferrous ions (Fe^{2+}) to hydroxyl radicals (·OH) according to the Fenton reaction Eq. (1). Superoxide anion also reacts with nitric oxide (NO), formed by nitric oxide synthases, to form the highly reactive peroxynitrite anion (-OONO), which reacts with H_2O_2 to form singlet oxygen according to Eq. (2), and this is only one of many possible mechanisms of formation of singlet oxygen in biological systems [4–6].

$$\mathrm{HOOH} + Fe^{2+} \rightarrow Fe^{3+} + {}^{-}\mathrm{OH} + HO^{\cdot} \quad (1)$$

$$\mathrm{ONOO}^{-} + \mathrm{HOOH} \rightarrow {}^{1}\mathrm{O}_2 + ONO^{-} + \mathrm{H_2O} \quad (2)$$

Lipid peroxidation involves a reaction between unsaturated lipids and oxygen. This may be enzyme-catalysed or non-enzymatic. Non-enzymatic lipid oxidation is either mediated by singlet oxygen, or it may involve free radical oxidation [7]. Singlet oxygen reacts by electrophilic addition to any of the double bonds in an unsaturated fatty acid such as linoleic acid (LA) to form hydroperoxide isomers such as the 10-, 12- and 13-LA hydroperoxides (10-LA-OOH, 12-LA-OOH and 13-LAOOH) as shown in **Figure 1**.

On the other hand, free radical oxidation begins by the abstraction of a hydrogen atom from a fatty acid, for example by the hydroxyl radical, to form a carbon centred radical, which rearranges to form a relatively stable conjugated radical (**Figure 2**). The latter reacts with oxygen to form a peroxyl radical, which abstracts a hydrogen from another fatty acid molecule to form a hydroperoxide and a new carbon centred radical, hence establishing a free radical chain reaction (**Figure 2**).

A fatty acid hydroperoxide can be converted to an alkoxyl radical by Fe^{2+} (**Figure 3**), in analogy to the conversion of H_2O_2 to the hydroxyl radical according to Eq. (1). The alkoxyl radical can be converted to a number of non-aldehydic products, including a hydroxy acid and a keto-acid, or it can cyclize to form an epoxy-allylic radical whose further oxygenation affords a hydroperoxy-epoxide

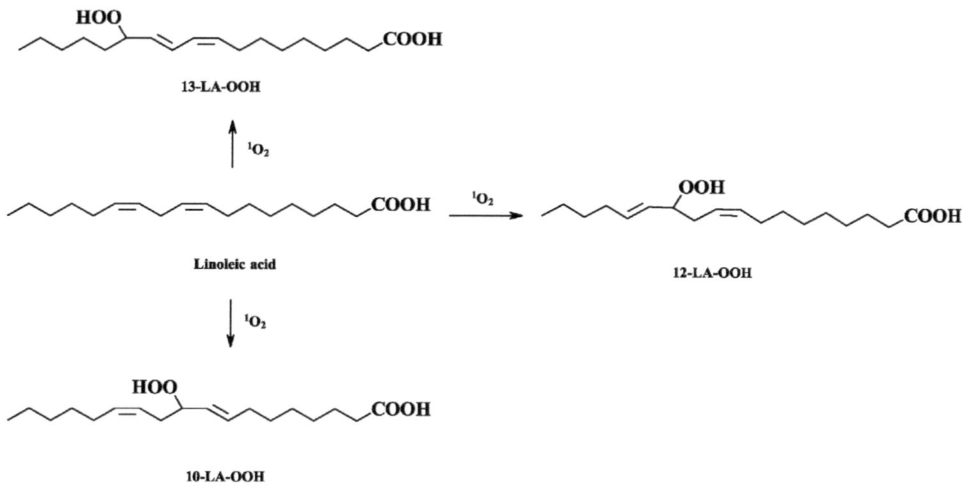

Figure 1.
Formation of different hydroperoxide isomers by the singlet oxygen-mediated oxidation of linoleic acid.

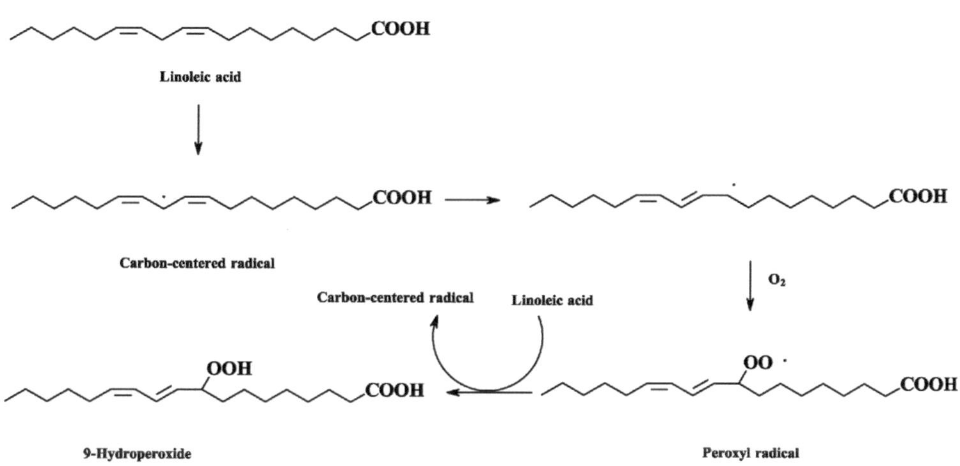

Figure 2.
*Free radical peroxidation of linoleic acid showing the formation of one of the two most readily formed hydroperoxides, the 9-hydroperoxide. The other easily formed hydroperoxide is the 13-hydroperoxide (shown in **Figure 1**).*

(not shown) that can further be converted to various products including epoxy-keto-acids, such as 12,13-epoxy-9-keto-10*E*-octadecenoic acid (**Figure 3**), which contributes to hypertension as discussed in Section 3.2.

An alkoxyl radical can also undergo beta scission (C-C cleavage) to form an aldehyde and a carbon centred radical, and this is only facile if the latter is a resonance stabilized allylic radical, such as would be formed from the 10-LA-OOH (**Figure 4**) or 12-LA-OOH but not 13-LA-OOH [8]. Beta scission is also facile if the carbon bearing the alkoxyl radical occurs next to another oxygen-bearing carbon [9]. Various pathways fulfilling these conditions have been proposed for the formation of the major bioactive lipid-derived aldehydes such as MDA, HNE, acrolein and glyoxal [9, 10].

Acrolein is mainly formed from PUFAS with more than two double bonds, such as arachidonic acid, eicosapentaenoic acid (EPA), docosapentaenoic acid (DPA) and docosahexaenoic acid (DHA) [9]. **Figure 5** shows examples of how MDA, HNE and glyoxal can all be formed from linoleic acid, the most abundant PUFA in most human tissues [3–6]. It starts with the 13-LA-OOH, formed by singlet

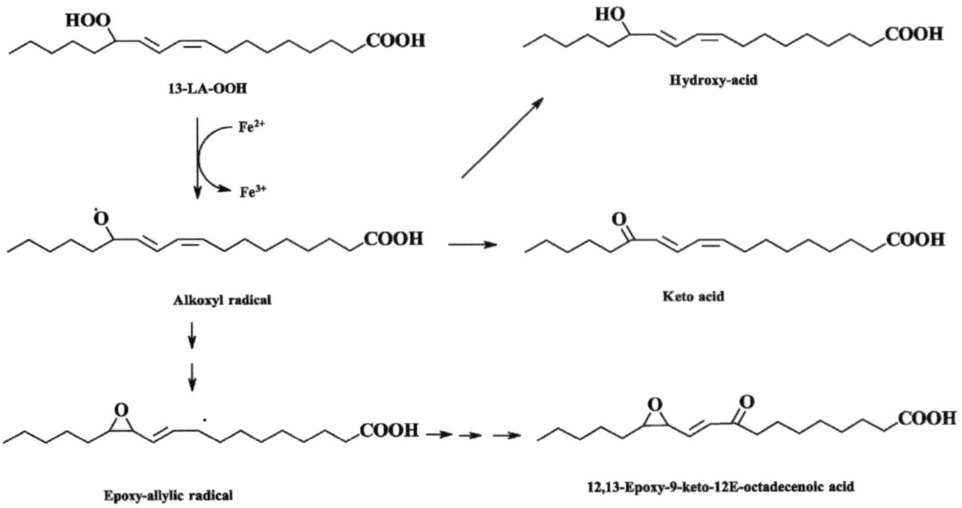

Figure 3.
Conversion of the 13-hydroperoxide of linoleic acid (13-LA-OOH) via the corresponding alkoxyl radical to different types of non-aldehydic products.

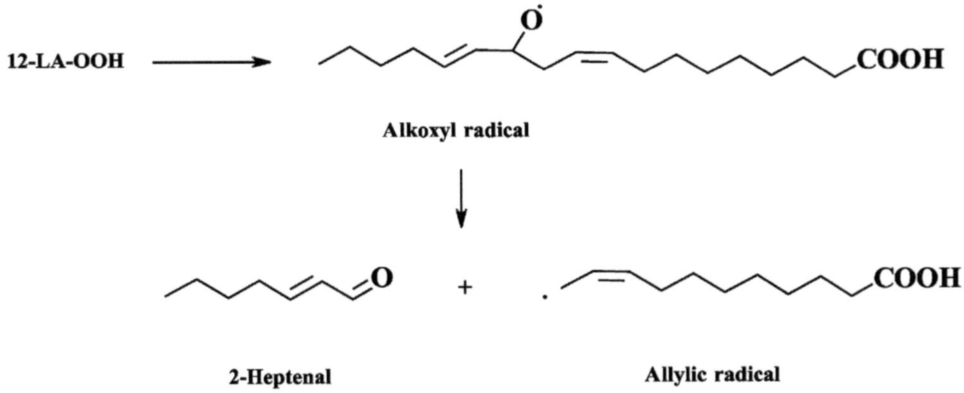

Figure 4.
b-Scission of an alkoxy radical to form an aldehyde (2-heptenal) and an allylic radical. Scission on the other side of the alkoxyl radical to form a vinyl radical and 12-oxo-9-dodecenoic acid is energetically unfavourable.

oxygen -mediated or free radical oxidation, which further reacts with singlet oxygen to form a hydroperoxy-dioxetane (addition of singlet oxygen to a conjugated double bond system forms dioxetanes rather than hydroperoxides). The dioxetane is unstable, and decomposes to form two aldehydes, namely 9-oxononanoic acid and 4-hydroperoxy-2-nonenal (4-HPNE). The former is one of the predominant products of linoleic acid oxidation, and contributes to hypertension through activating phospholipase A2 (Section 3.2). A primary amine (RNH_2) such as lysine or phosphatidylethanolamine may catalyse the conversion of 4-HPNE via a dioxetanyl anion to a dioxetane whose cleavage affords MDA and hexanal (**Figure 5**). While it has long been known that linoleic acid is a precursor of MDA, albeit not as readily as from more highly unsaturated PUFAS, its mechanism of formation from linoleic acid remained elusive [11]. 4-HPNE can alternatively react with another singlet oxygen molecule to form a hydroperoxy-dioxetanyl aldehyde whose decomposition affords glyoxal and 2-hydroperoxy-heptanal (not shown). 4-HPNE can also be converted to an alkoxyl radical, which can abstract a hydrogen to form 4-HNE, or to an epoxy-alkyl radical which can rearrange to an ether radical whose further

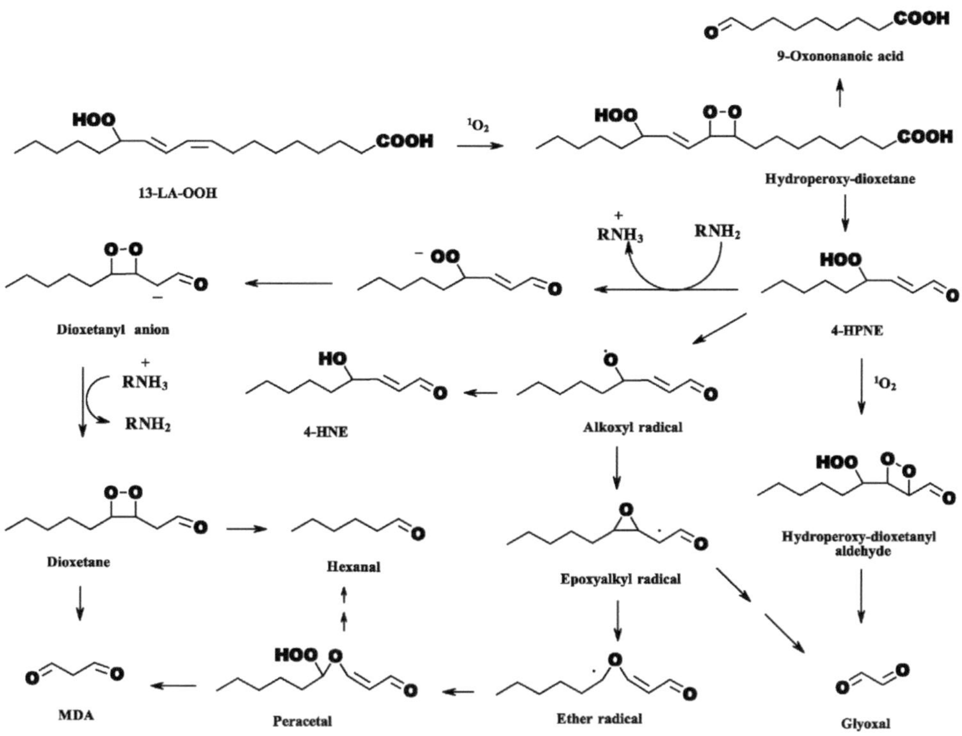

Figure 5.
Mechanisms of the conversion of the 13-hydroperoxide of linoleic acid (LA-OOH) to 4-HNE, MDA, glyoxal, hexanal and 9-oxo-nonanoic acid. Other pathways to these products exist but are not shown.

reaction with oxygen leads to formation of a per acetal that can decompose to MDA and hexane. The epoxy-alkyl radical can also directly react with oxygen to form a hydroperoxy-epoxide whose decomposition affords glyoxal. In the cell, glutathione peroxidase may contribute to the conversion of 4-HPNE to 4-HNE.

3. Lipid peroxidation products contribute to the development of the metabolic syndrome

Lipid oxidation products influence the pathogenesis of metabolic syndrome components such as obesity, hypertension, impaired fasting glucose/diabetes, and dyslipidemia, in various ways [12].

3.1 Obesity

Obesity occurs when adipocytes increase in number and/or size, coupled with increased fat storage and reduced fat oxidation. Adipose tissue (AT) is functionally classified as brown or white (BAT and WAT, respectively). BAT consists of adipocytes specialized for thermogenesis, and hence contribute to reduction of obesity; while WAT, the major type of adipose tissue in humans, has less capacity for fat oxidation, and may contribute to obesity [13]. White adipocytes can exist in or acquire a brown-like (beige or brite) phenotype with higher fat oxidation than ordinary white adipocytes, and a higher number of beige adipocytes reduces an individual's susceptibility to obesity [13]. Expansion of WAT by differentiation of preadipocytes (hyperplasia) into mature adipocytes with adequate lipid filling and fat oxidation capacity is beneficial for safe storage of fat; but mere expansion of mature adipocytes because of excessive lipid filling and reduced fat oxidation (hypertrophy) is associated with adverse health outcomes.

A certain amount of ROS is required for proper preadipocyte and mature adipocyte physiology. However, oxidative stress and excessive autophagy may inhibit preadipocyte differentiation and promote hypertrophy of mature adipocytes (**Figure 6**) [14]. Likewise, brown or beige adipocytes have many mitochondria for the enhanced fat oxidation, but mitochondrial oxidative stress causes loss of the mitochondria through mitophagy, thus leading to whitening, increased lipid storage and hypertrophy (**Figure 6**) [15]. Adenosine 5-monophosphate kinase (AMPK), sirtuins 1 and 3, protein kinase B (akt), peroxisome proliferator activated receptor gamma and alpha (PPARγ and PPARα, respectively), and PPARγ coactivator-1α (PGC-1α) are among the proteins that reduce oxidative stress and/or promote mitochondrial biogenesis in adipocytes [16, 17]. Both protein kinase A (PKA) and akt are required for PPARγ expression [18], which is required for differentiation of both brown and white adipocytes [19]. PPARγ promotes thermogenesis in mature brown adipocytes through activation of uncoupled protein 1 (UCP-1), and by upregulating glycerol kinase which catalyses glycerol-3-phosphate synthesis, which is required for TG synthesis [20]. While this looks paradoxical, TG synthesis may help reduce the lipotoxicity and oxidative stress induction by free fatty acids (discussed in Section 4), and allow targeted, β-adrenergic signaling-associated release of fatty acids for mitochondrial oxidation. AMPK activates autophagy and induces the transcription factor nrf2; and the latter upregulates antioxidant enzymes such as catalase, glutathione peroxidase, superoxide dismutase and heme oxygenase 1 [21]. Sirt1, which is mainly localized in the nucleus, increases the expression catalase and SOD as reviewed by Iside et al. [22]. In addition, it upregulates autophagy genes, and autophagy defect associated with its inhibition promotes release of exosomes which induce toll-like receptor 4 (TLR4) signaling, downstream activation of nuclear factor kappa B (NF-kB), and NF-kB-mediated upregulation of oxidative stress and inflammation-promoting genes [23].

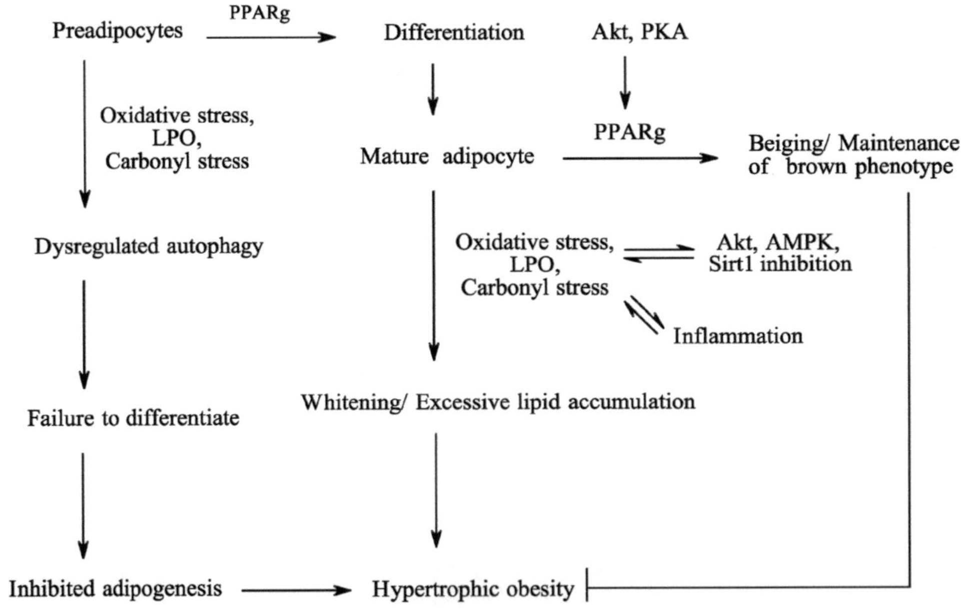

Figure 6.
Role of oxidative stress and lipid oxidation-induced carbonyl stress in the pathogenesis of hypertrophic obesity.

Conditions that promote adipose tissue oxidative stress, including inappropriate diets (Section 4), induce lipid oxidation, and the latter generates carbonyl stress due to formation of various aldehydes as explained in Section 2. These aldehydes, including HNE and acrolein modify and inhibit AMPK and sirt1, thus amplifying oxidative stress and their own formation (**Figure 6**) [24–26]. HNE also carbonylates insulin receptor substrate1/2 (IRS1), leading to degradation and inhibition of the latter, thus inducing insulin resistance and downstream akt inhibition [27]. Thus, insulin resistant obese individuals have lower akt, AMPK ad sirt1 activity, but higher reactive carbonyls and carbonylated proteins [28]. Acrolein and HNE additionally aggravate oxidative stress through readily reacting with, and depleting the antioxidant glutathione [29, 30]. They modify the endoplasmic reticulum calcium pump SERCA, leading to its inhibition and ER stress [3], which aggravates oxidative stress, insulin resistance, sirt1 inhibition, expression of the pro-inflammatory cytokines, TNFα and IL6, and adipocyte whitening [31, 32]. Glutathionylated HNE and other aldehydes released from adipocytes under conditions of oxidative stress promote macrophage infiltration into WAT, and their acquisition of a proinflammatory M1 phenotype [33]. Malondialdehyde reacts with albumin, and the MDA-albumin conjugates promote a proinflammatory phenotype in macrophages and T cells [34]. Cytokines such as interleukin1-β, released from inflammatory macrophages, in turn promote adipocyte oxidative stress and whitening [35].

3.2 Hypertension

Arterial hypertension occurs because of (i) increased renal retention of sodium and water (ii) dysregulation of vasodilators and vasoconstrictors and (iii) arterial stiffness. Obesity is a major risk factor for hypertension [36]. For example, adiponectin inhibits adrenal production of aldosterone, a potent inducer of hypertension [37], but obesity reduces adiponectin secretion and increases circulating aldosterone [38]. Thus, by promoting obesity, lipid peroxidation products indirectly promote hypertension. However, they also induce hypertension independently of obesity. For example, the non-aldehydic linoleic acid oxidation product, 12,13-epoxy-9-keto-10-*E*-octadecenoic acid (Shown in **Figure 3**) also promotes adrenal production of aldosterone to induce hypertension [39].

Aldosterone binds to the renal tubular epithelial cell mineralocorticoid receptor, which, as a transcription factor, upregulates the expression of the epithelial sodium channel, which promotes sodium retention [40]. Independently of gene transcription, aldosterone activates the non-receptor tyrosine kinase cSrc in these cells, probably through the angiotensin receptor type 1 (AT1R), and cSrc activates epidermal growth factor receptor (EGFR) signaling, leading to activation of the mitogen activated protein kinase Erk1/2 [40]. Erk1/2 activates Na^+/K^+ ATPase, which promotes sodium and water retention [41]. Aldosterone-cSrc signaling also induces oxidative stress [40], which induces formation of lipid oxidation products. 4-HNE, inhibits AMPK and sirt1, thus inhibiting eNOS, leading to reduced NO bioavailability, and this causes increased transactivation of EGFR and downstream Erk1/2 [42–44]. Thus, blood HNE levels are increased in hypertension [45], and the latter can be ameliorated by carbonyl quenching [46]. Oxidized low density lipoprotein, which contains oxidatively modified lipids such as HNE, induces oxidative stress in renal tubular endothelial cells, which activates the renal renin-angiotensin system (RAS); whose component, angiotensin 2, overstimulates sodium transporters and thus induces hypertension [47, 48]. Hypertension in turn promotes oxidative stress and LDL oxidation, thereby creating a vicious circle [49].

Inhibition of endothelial cell sirt1, sirt3 and AMPK, which can be mediated by LPO products, causes inhibition of endothelial nitric oxide synthase (eNOS) and decreases production of NO, the main arterial vasodilator [50–52]. HNE induces endothelial cell insulin resistance, and the associated akt inhibition both inhibits eNOS and upregulates the vasoconstrictor, endothelin [3, 53]. The dysfunctional endothelial cells further produce pro-inflammatory factors such as TNFα, IL-1β, IL-8 and MCP-1 which recruit circulating neutrophils, platelets and monocytes, and the latter differentiate into macrophages [53, 54]. Neutrophils, monocytes and macrophages secrete myeloperoxidase [55]. Myeloperoxidase oxidizes LDL [56]. It also promotes the activation of endothelial cyp4a12a, which catalyzes the formation of 20-hydroxy-eicosatetraenoic acid (20-HETE) from arachidonic acid [57]. 20-HETE upregulates endothelial RAS components including angiotensin 2, a potent vasoconstrictor, which also induces aldosterone secretion [58]. Both angiotensin 2 and aldosterone aggravate endothelial oxidative stress and dysfunction. Androgens promote 20-HETE synthesis, and this may explain the higher occurrence of hypertension in men than women [58].

Stiffness of the coronary artery and other major arteries inhibits their systolic dilatation, and thus promotes systolic hypertension [59]. Degradation of the elastic fiber, elastin, in the walls of the major arteries, and its replacement with collagen fibres is a hallmark of the pathogenesis of arterial stiffness [59]. The myeloperoxidase product, 20-HETE, activates matrix metalloproteinase 12 (MMP-12, macrophage elastase), which degrades elastin [60]. Myeloperoxidase additionally inhibits the elastase inhibitor, α1, and this is antagonized by sulfur compounds such as glutathione [61]. Acrolein and HNE, on the other hand, deplete glutathione [62]. 20-HETE additionally sensitizes vascular smooth muscle cells to stimuli that promote their dedifferentiation and proliferation [58], which contributes to arterial stiffening especially of the muscular arteries [63]. One of the most readily formed aldehydic linoleic acid oxidation products, 9-oxononanoic acid (**Figure 3**) activates phospholipase A2 (PLA_2) leading to generation of eicosanoids and thromboxane A2 in blood [64]. Thromboxane A_2 causes vasoconstriction and the proliferation of smooth muscle cells [65].

Malondialdehyde forms collagen crosslinks that prevent collagen degradation, thus promoting arterial stiffness [66]. Thus, MDA-modified LDL independently predicts arterial stiffness [67]. Glyoxal contributes to arterial stiffness by reacting with collagen to form advanced glycation end products such as GOLA, GOLD, GODIC and carboxymethyl lysine (CML) [68]. CML induces endothelial oxidative stress through the RAGE receptor, which activates components of NF-kB signaling that promote expression of collagen 1 and 2 [69, 70].

3.3 Dyslipidemia

Dyslipidemia in metabolic syndrome is defined by elevated circulating triglycerides (hypertriglyceridemia) or low levels of high-density lipoprotein cholesterol (low HDLc); and hepatic steatosis, a component of non-alcoholic fatty liver disease (NAFLD) is its main risk factor [71]. This is because, in hepatic steatosis, there occurs greater production and secretion of triglyceride-rich very low-density lipoproteins (VLDL), leading to hypertriglyceridemia; as well as higher hepatic lipase activity, which increases the hepatic uptake and degradation of HDL [71]. Hepatocyte oxidative stress, ER stress and associated lipid peroxidation are involved in the development of hepatic steatosis [72, 73], and this makes lipid peroxidation an important factor in the development of dyslipidemia [74].

Low HDLc also occurs in obesity independently of elevated triglycerides, indicating that it occurs even independently of NAFLD [75]. Hypoadiponectinemia, which depends on obesity rather than NAFLD [76], may cause reduced HDLc through increased hepatic lipase activity; reduced hepatic expression of the HDL protein apo A; reduced expression of the cholesterol export protein ABCA1 which transfers cholesterol to HDL; and upregulated synthesis of LCAT which transfers cholesterol from HDLc to chylomicrons [77]. Obesity is also associated with increased plasma TNFα [78] which suppresses hepatocyte apo AI gene expression via ERK and JNK [79]. HNE contributes to JNK over-activation in hepatocytes [80].

3.4 Prediabetes and diabetes

Diabetes is a state of elevated postprandial and/or fasting blood glucose that, if not controlled, leads to the damage of various organs; while prediabetes refers to an intermediate level of fasting and/or postprandial blood glucose, higher than normal but less than diabetic blood glucose levels [1]. It is an earlier stage toward the development of diabetes, but which can revert to normoglycemia. The major causes of (pre)diabetes are pancreatic alpha and beta cell dysfunctions leading to glucagon over-secretion and insulin under-secretion, respectively; coupled with skeletal muscle, adipose and/or hepatic insulin resistance [81].

Both obesity and hypertension contribute to the pathogenesis of prediabetes, hence the lipid oxidation products that induce obesity and hypertension indirectly promote diabetes thereby. Nevertheless, lipid oxidation products also directly promote (pre)diabetes. For example, malondialdehyde was found to dose-dependently reduce the insulin content in the pancreas and to contribute to beta cell death [82]. HDL prevents beta cell apoptosis and diabetes by promoting cholesterol efflux from these cells, but acrolein- or MDA-modified HDL loses this protective property [83–85]. oxLDL impairs insulin gene expression and causes death of pancreatic beta cells, through induction of oxidative stress and ER stress [86]. As already discussed, lipid oxidation products induce endothelial dysfunction. Pancreatic endothelial cell dysfunction contributes to diabetes, being associated with leukocyte recruitment and increased production of proinflammatory cytokines [87]. Cytokines such as IL-1β and TNFα induce alpha and beta cell oxidative stress [88] and associated lipid peroxidation. Insulin resistance, which can be induced by HNE and acrolein, is part of the alpha cell and beta cell dysfunctions leading to diabetes [81].

4. Role of dietary constituents in inducing tissue oxidative stress, lipid peroxidation and the metabolic syndrome

Diets high in saturated fatty acids, cholesterol, sugar, salt, and red meat, contribute to higher lipid oxidation in the tissues and organs that have a central role in the metabolic syndrome, such as adipose tissue, endothelial tissue, muscle, liver and pancreas.

Although saturated fatty acids do not undergo peroxidation, they contribute to the induction of oxidative stress in cells, which then leads to peroxidation of unsaturated fatty acids. For example, the most abundant saturated fatty acid in the diet, palmitic acid, is a key substrate for the first reaction in ceramide biosynthesis [89]. Ceramides induce oxidative stress, for example by inhibiting components of the electron transport chain [90].

Palmitate also induces oxidative stress and ER stress independently of ceramide. For example, it increases diacylglycerol levels, which is associated with activation of protein kinase C (PKC), which inhibits the Kreb's cycle enzymes aconitase and isocitrate dehydrogenase [91]. Thus, the acetyl COA generated from peroxisomal and mitochondrial fatty acid beta oxidation accumulates in the cell, promoting acetylation of mammalian target of rapamycin complex (MTORC-1) and high mobility group box-1 (HMGB-1), as has been demonstrated in hepatocytes [92]. Acetylation activates MTORC-1, which inhibits akt and further promotes oxidative stress by upregulating the expression of TLR4, thus upregulating the NF-kB-NADPH oxidase/iNOS axis [92]. Acetylation of HMGB-1 causes its translocation out of the cell, enabling it to induce oxidative stress by interacting with the receptor for advanced glycation end products (RAGE) as well as TLR4, which both induce NF-kB activation [93]. Obesity is associated with increased circulating HMGB-1, which accelerates the pathogenesis of obesity, hypertension and diabetes [72, 73, 94, 95]. TLR2/4 signaling also activates RAS components including angiotensin 2, whose signaling via its receptor AT1R induces NFkB and oxidative stress [3, 96].

PUFAS undergo peroxidation during cooking as well as in the digestive tract [97]. This is more pronounced when they are part of a meal containing meat, especially red meat, which has higher myoglobin content; since iron from the latter promotes lipid oxidation [98]. This leads to a postprandial increase in circulating carbonyls such as malondialdehyde and HNE, which promote oxidative stress, HDL modification and postprandial inflammation [98, 99]. On the other hand, absorbed, unoxidized unsaturated fatty acids including MUFAs and PUFAs reduce palmitate-induced oxidative stress and lipotoxicity in many cell types by promoting the incorporation of palmitate into TGs for safe storage [100–102]. Nevertheless, high concentrations of arachidonic acid also induce deleterious effects. Thus, supplementation of arachidonic acid to a high fat diet led to enhanced obesity in mice [103], which is attributable to the fact that this n-6 fatty acid promotes adipogenesis from preadipocytes, but its cyclooxygenase-mediated oxidation products, prostaglandins E2 and F2a (PGE2 and PGF2a) inhibit browning via ERK activation and associated decrease in UCP-1 expression [104, 105]. These prostaglandins activate NF-kB, diminish adiponectin production, upregulate pro-inflammatory mediators such as TNFα and MCP-1, and thus promote macrophage activation [106]. They promote oxidative stress and lipid oxidation, and the lipid oxidation product HNE in turn induces cyclooxygenase 2 [107]. Adipose inflammation has systemic effects, hence adipose tissue arachidonic acid was found to be independently associated with abdominal obesity, dyslipidemia, hypertension and fasting glucose [108]. Its myeloperoxidase products, 20-HETE is associated with insulin resistance and hyperglycemia [109]. Since humans synthesize arachidonic acid from linoleic acid, the arachidonic acid content in human adipose tissues does not necessarily reflect its dietary intake [110].

Although linoleic acid is a precursor of arachidonic acid, studies of its effects on the metabolic syndrome have given mixed results, with both harmful and protective roles reported [111–113]. The differences are partly due to genetic factors. For example, there are individual and ethnic differences in the expression of fatty acid desaturase 1 and 2 (FADS 1/2); with genotypes favouring greater FADS1/2 activity and arachidonic acid synthesis being associated with greater susceptibility to metabolic dysregulation [114, 115]. Black people and Indians significantly generate arachidonic acid from dietary linoleic acid, unlike people of European origin [114, 116]. A high adipose tissue linoleic: arachidonic acid is inversely associated with cardiovascular mortality and hypertension [112]. Likewise, a low linoleic: arachidonic acid ratio in plasma phospholipids is associated with hypertension [117]. Polymorphisms in the receptor for oxLDL, Lox-1, might also determine differences in the response to increased dietary linoleic acid; since this PUFA increases Lox-1 expression in aortic endothelial cells [118]. The effects of linoleic acid may also be dependent on the overall diet. If the diet is high in other factors that induce oxidative stress such as dietary sugar and salt, the pro-oxidative environment thus created may abrogate potential linoleic acid benefits through its increased oxidation. This is in analogy to the fact that high glycemic index foods abrogate the anti-obesity effects of fish oil [119]. A high linoleic acid diet may also be unfavourable for people who have already developed some component of the metabolic syndrome and thus have a more pro-inflammatory status.

Oleic acid is the major dietary fatty acid in the Mediterranean diet, which is generally associated with health benefits. This fatty acid is relatively resistant to peroxidation. Besides promoting the safe storage of palmitate in TGs, it induces thermogenesis by upregulating adipose triglyceride lipase and hormone sensitive lipase, which induce lipolysis coupled with fatty acid oxidation [120]. It promotes M2 macrophage phenotype in visceral adipose [121].

The dietary n-3 fatty acids are generally highly susceptible to oxidation because they all contain at least 3 double bonds. Although they are not 4-HNE precursors, decomposition of their hydroperoxides very readily produces acrolein, MDA, glyoxal and methylglyoxal. Despite this, they are largely beneficial, suppressing development of the metabolic syndrome [122]. In adipocytes, their binding to the GPR120/Ffar4 receptor inhibits TLR2 and TLR4 signaling and associated NFkB activation, oxidative stress and inflammation [123]. This receptor also upregulates miR-30b and 378, and induces FGF21 secretion, whose signaling activates AMPK, promotes browning and induces adiponectin [123–126]. The n-3 PUFAs are metabolized by cyclooxygenase to resolvins, protectins, maresins and isoprostanes which help in resolving inflammation [127].

A high dietary n-3: n-6 PUFA ratio has been found to be protective against the metabolic syndrome in some studies but not others [128, 129]. This might be partly due to inter-individual differences in the metabolism of n-6 fatty acids.

High carbohydrate diets promote obesity because excess sugars are stored as lipids. High sucrose or high fructose diets are particularly obesogenic [130]. Fructose metabolism robustly increases palmitate synthesis in adipocytes [131]. Moreover, fructose metabolism is associated with decreased cellular ATP, purine degradation and activation of xanthine oxidase which generates reactive oxygen species and associated lipid peroxidation [132], which is involved in adipocyte whitening and less thermogenesis. Uric acid, a product of purine degradation also induces oxidative stress through increased NADPH oxidase activity and RAS activation [131–134].

High salt (sodium chloride) diets promote obesity, by salt-induced activation of adipocyte Na/K+ ATPase, which is coupled to activation of src, which generates ROS, and transactivates PI3-K-Akt–MTOR and EGFR-ERK/MAPK pathways [135, 136]. This is associated with increased expression of proinflammatory mediators such as TNFα, MCP-1, COX-2, IL-17A, IL-6, leptin, and leptin [136, 137]. Sodium chloride also activates Na+/K+ ATPase and induces oxidative stress in endothelial cells and renal tubular epithelial cells, thereby promoting hypertension [132], and this is also subject to genetic susceptibility [138].

High dietary cholesterol is associated with a high risk of dyslipidemia [139]. Cholesterol-rich chylomicron remnants mainly deliver their cholesterol to the liver, and cholesterol accumulation in hepatocytes strongly induces oxidative stress, by modification of the mitochondrial membrane and limiting import of glutathione into the mitochondria, as well by inducing ER stress and proinflammatory cytokines [140].

The lipopolysaccharide (LPS) component of the walls of gram-negative bacteria is a pro-inflammatory molecule that contributes to metabolic low-grade inflammation (endotoxemia), by signaling through TLR2 and 4 in various cell types, leading to NFkB activation and release of pro-inflammatory cytokines. High sucrose and high saturated fat diets promote the growth of gram-negative bacteria, and thus increase the entry of LPS into the circulation [141].

5. Mechanisms of the antioxidant and metabolic syndrome-suppressing effects of dietary factors

While some food components promote a pro-oxidative and pro-inflammatory state as discussed in the previous section, other dietary factors inhibit oxidative stress and inflammation. They do this through various mechanisms, but the most widely considered mechanisms are those associated with lipid oxidation, including scavenging of free radicals such as peroxyl radicals and alkoxyl radicals, chelation of metal ions that participate in formation of such radicals, and singlet oxygen quenching.

5.1 Free radical scavenging and singlet oxygen quenching

Carotenoids, phenolic substances, tocopherols and ascorbic acid are well known for their antioxidant activities targeting the neutralization of reactive radicals and/or singlet oxygen quenching. Thus, carotenoids reduce oxidative stress and lipid oxidation, resulting in adipocyte beiging and obesity prevention [142]. There is decreased adipose beta carotene in obese subjects, and this was suggested to at least partly be due to their depletion under the high ROS environment [142]. Likewise, tocopherols and tocotrienols have been shown to be protective against all components of the metabolic syndrome [143]. Thus, the high tocotrienol content of palm oil may reduce its potential harm from the high palmitate content [144]. Unfortunately, radical scavenging antioxidants also exhibit pro-oxidant activity, depending on their concentrations and the level of prooxidative factors [145]. Hence, there is need to consider a broad range of dietary factors that prevent oxidative by alternative mechanisms, such as those outlined hereafter. A single molecule can act by multiple mechanisms, and the more mechanisms involved, the greater might be the benefit.

5.2 Insulin-mimicking

Insulin signaling activates akt, which reduces oxidative stress by promoting mitophagy and by activating nrf2 to induce antioxidant enzymes [81, 146]. Moreover, nrf2, via heme oxygenase 1 (HO-1), inhibits NFkB and associated upregulation of NADPH oxidase and iNOS [147]. Quercetin and ferulic acid are examples of molecules that have demonstrated oxidative stress and metabolic syndrome amelioration at least partly through PI3K-akt signaling in various cell types [148–150]. Resveratrol and ferulic acid inhibit LPS- and oxidative stress-induced intestinal barrier injury through this signaling pathway [151, 152].

5.3 AMPK and SIRT1 activation

AMPK and/or sirt1 reduce mitochondrial oxidative stress in adipocytes, pancreatic beta cells, hepatocytes, endothelial cells, and thus are useful in preventing all aspects of the metabolic syndrome. In addition to insulin mimicking, quercetin and ferulic acid, also activate these proteins [153–155].

5.4 Adiponectin and adiponectin receptor enhancement

Compounds that activate AMPK, sirt1 and/or PI3K-akt in adipose tissues limit adipocyte hypertrophy and inflammation, and enhance adiponectin production. Adiponectin has systemic effects in reducing insulin resistance and oxidative stress, because it activates both PI3K-akt and AMPK in insulin target tissues, and also promotes anti-inflammatory polarization of macrophages [156]. Dietary compounds than ameliorate metabolic syndrome through enhanced adiponectin secretion and/or upregulating adiponectin receptor include n3-fatty acids, sesamin, the citrus derived polymethoxyflavonoids nobiletin and tangeretin, quercetin and resveratrol [157–160].

5.5 Ceramide reduction

Adiponectin signaling increases ceramidase activity, thus reducing ceramide levels [161]. Hence, the adiponectin and adiponectin receptor enhancers should contribute to reducing ceramide-induced oxidative stress. Not much research has been done along this line, but it has been reported that DHA inhibits ceramide biosynthesis [162]. In mice, dietary inulin reduces ceramide synthesis by suppressing neutral sphingomyelinase expression and activity [163].

5.6 Vasodilation

Vasodilation reduces blood pressure, and thus reduces pressure-dependent oxidative stress as well as LDL oxidation and Lox-1 dependent oxidative stress [164]. Thus, for people with prehypertension or hypertension, vasodilation may be a major strategy for reducing oxidative stress and lipid oxidation. Cinnamaldehyde has vasodilatory and antihypertensive activity through effects on smooth muscle contractility [165]. Dietary nitrate achieves vasodilation through NO release, and this is associated not only with pressure regulation, but also other components of the metabolic syndrome including blood glucose and lipid improvement [166]. Adiponectin induces AMPK dependent eNOS activation in endothelial cells, hence adiponectin enhancers such as imperatorin also promote NO synthesis and vasodilatation [167].

5.7 Reactive carbonyl, ALEs and AGEs scavenging

Scavengers of reactive carbonyls such as HNE, acrolein and MDA have been demonstrated to ameliorate oxidative stress, lipid peroxidation and the metabolic syndrome. Examples of compounds with such effects include carnosine, carnosinol, epigallocatechin-3-gallate and the mulberry anthocyanins cyanidin 3-glucoside (C3G) and cyanidin 3-rutinoside (C3R) [46, 168–170]. Aminoguanidine attenuates hypertension by scavenging AGES [171].

5.8 Gut microbiota modulation

Probiotic microorganisms suppress the growth of pathogenic microorganisms. They also produce metabolites such as short chain fatty acids with beneficial effects on the metabolic syndrome. For example, butyrate promotes PI3K-akt signaling to prevent oxidative stress and maintain intestinal barrier integrity [172, 173]. Quercetin, resveratrol and n-3 fatty acids have been demonstrated to positively influence gut microbiota and decrease intestinal barrier permeability in animal studies [153, 174].

6. Conclusions

Lipid peroxidation is a major contributor to the pathogenesis of the metabolic syndrome, especially through highly reactive and bioactive aldehydes such as acrolein, 4-hydroxy-2-nonenal, malondialdehyde and glyoxal. Mechanisms of formation of these products are now well-understood. For example, this article has highlighted that formation of MDA from linoleic acid may be easier than previously thought. The mentioned aldehydes propagate oxidative stress and inflammation by inducing insulin resistance, inhibiting sirt1 and AMPK, reducing adiponectin secretion, as well as forming AGEs and ALEs that activate the RAGE receptor. Inhibiting LPO and the LPO product-associated oxidative stress and inflammation is necessary for preventing and/or ameliorating progression of the metabolic syndrome. This may not be effectively accomplished by dietary agents that merely scavenge free radicals and/or quench singlet oxygen, but also by those that inhibit the signaling pathways that generate non-lipid ROS, or scavenge the reactive carbonyls, ALEs and AGEs. In addition, saturated fat, sugar, meat, and salt, that fuel the signaling pathways that initiate LPO should be reduced. The metabolic influence of some dietary components such as salt and n-6 PUFAs is particularly influenced by genetics, and this should be duly considered when making dietary recommendations.

Author details

Arnold N. Onyango
School of Food and Nutrition Sciences, Jomo Kenyatta University of Agriculture and Technology, Nairobi, Kenya

*Address all correspondence to: arnold.onyango@jkuat.ac.ke

References

[1] Alberti, K. G. M.M, Zimmet, P.Z. for the WHO Consultation (1998). Definition, diagnosis and classification of diabetes mellitus and its complications. Part 1: diagnosis and classification of diabetes mellitus. Provisional report of a WHO Consultation. Diabetic Medicine 1998; 15: 539-553

[2] National Cholesterol Education Program (NCEP): Expert Panel on Detection and Treatment of High Blood Cholesterol in Adults (2002). Third Report of the National Cholesterol Education Program (NCEP) Expert Panel on Detection, Evaluation, and Treatment of High Blood Cholesterol in Adults (Adult Treatment Panel III) final report. Circulation 106, 3143-3421

[3] Onyango, A. N. (2018). Cellular stresses and stress responses in the pathogenesis of insulin resistance. *Oxidative Medicine and Cellular Longevity*, Article ID 4321714.

[4] Onyango, A. N. (2016). Endogenous generation of singlet oxygen and ozone in human and animal tissues: mechanisms, biological significance, and influence of dietary components. Oxidative Medicine and Cellular Longevity, *2016*.

[5] Onyango, A. N. (2016). Formation of aldehydic phosphatidylcholines during the anaerobic decomposition of a phosphatidylcholine bearing the 9-hydroperoxide of linoleic acid. *BioMed Research International*, Article ID 8218439.

[6] Onyango, A. N. (2017). The contribution of singlet oxygen to insulin resistance. *Oxidative Medicine and Cellular Longevity*, Article ID 8765972.

[7] Girotti, A. W. (1985). Mechanisms of lipid peroxidation. Journal of Free Radicals in Biology & Medicine, *1*(2), 87-95.

[8] Onyango, A. N., Inoue, T., Nakajima, S., Baba, N., Kaneko, T., Matsuo, M., & Shimizu, S. (2001). Synthesis and stability of phosphatidylcholines bearing polyenoic acid hydroperoxides at the sn-2 position. Angewandte Chemie International Edition, *40*(9), 1755-1757.

[9] Onyango, A. N. (2012). Small reactive carbonyl compounds as tissue lipid oxidation products; and the mechanisms of their formation thereby. Chemistry and physics of lipids, *165*(7), 777-786.

[10] Schneider, C., Porter, N. A., & Brash, A. R. (2008). Routes to 4-hydroxynonenal: fundamental issues in the mechanisms of lipid peroxidation. Journal of Biological Chemistry, *283*(23), 15539-15543.

[11] Onyango, A. N., & Baba, N. (2010). New hypotheses on the pathways of formation of malondialdehyde and isofurans. Free Radical Biology and Medicine, *49*(10), 1594-1600.

[12] Pillon, N. J., & Soulage, C. O. (2012). Lipid peroxidation by-products and the metabolic syndrome. Lipid peroxidation, 409-436.

[13] Zoico, E., Rubele, S., De Caro, A., Nori, N., Mazzali, G., Fantin, F., ... & Zamboni, M. (2019). Brown and beige adipose tissue and aging. Frontiers in endocrinology, *10*, 368.

[14] Castro, J. P., Grune, T., & Speckmann, B. (2016). The two faces of reactive oxygen species (ROS) in adipocyte function and dysfunction. Biological Chemistry, *397*(8), 709-724.

[15] Cedikova, M., Kripnerová, M., Dvorakova, J., Pitule, P., Grundmanova, M., Babuska, V., ... & Kuncova, J. (2016). Mitochondria in white, brown, and beige adipocytes. *Stem cells international*, *2016*.

[16] Andrade, J. M. O., Barcala-Jorge, A. S., Batista-Jorge, G. C., Paraíso, A. F., de Freitas, K. M., de Farias Lelis, D., ... & Santos, S. H. S. (2019). Effect of resveratrol on expression of genes involved thermogenesis in mice and humans. Biomedicine & Pharmacotherapy, *112*, 108634.

[17] Bargut, T. C. L., Souza-Mello, V., Aguila, M. B., & Mandarim-de-Lacerda, C. A. (2017). Browning of white adipose tissue: lessons from experimental models. Hormone molecular biology and clinical investigation, *31*(1).

[18] Kim, S. P., Ha, J. M., Yun, S. J., Kim, E. K., Chung, S. W., Hong, K. W., ... & Bae, S. S. (2010). Transcriptional activation of peroxisome proliferator-activated receptor-γ requires activation of both protein kinase A and Akt during adipocyte differentiation. Biochemical and Biophysical Research Communications, *399*(1), 55-59.

[19] Sanchez-Gurmaches, J., Calejman, C. M., Jung, S. M., Li, H., & Guertin, D. A. (2019). Brown fat organogenesis and maintenance requires AKT1 and AKT2. Molecular Metabolism, *23*, 60-74.

[20] Lasar, D., Rosenwald, M., Kiehlmann, E., Balaz, M., Tall, B., Opitz, L., ... & Wolfrum, C. (2018). Peroxisome proliferator activated receptor gamma controls mature brown adipocyte inducibility through glycerol kinase. Cell Reports, *22*(3), 760-773.

[21] Xu, Q., Fan, Y., Loor, J. J., Liang, Y., Sun, X., Jia, H., ... & Xu, C. (2021). Adenosine 5′-monophosphate-activated protein kinase ameliorates bovine adipocyte oxidative stress by inducing antioxidant responses and autophagy. Journal of Dairy Science.

[22] Iside, C., Scafuro, M., Nebbioso, A., & Altucci, L. (2020). SIRT1 activation by natural phytochemicals: an overview. Frontiers in Pharmacology, *11*, 1225.

[23] Li, F., Li, H., Jin, X., Zhang, Y., Kang, X., Zhang, Z., ... & Sun, H. (2019). Adipose-specific knockdown of Sirt1 results in obesity and insulin resistance by promoting exosomes release. Cell Cycle, *18*(17), 2067-2082.

[24] Lu, Q., Mundy, M., Chambers, E., Lange, T., Newton, J., Borgas, D., ... & Rounds, S. (2017). Alda-1 protects against acrolein-induced acute lung injury and endothelial barrier dysfunction. American Journal of Respiratory Cell and Molecular Biology, 57(6), 662-673.

[25] Shearn, C. T., Backos, D. S., Orlicky, D. J., Smathers-McCullough, R. L., & Petersen, D. R. (2014). Identification of 5′ AMP-activated kinase as a target of reactive aldehydes during chronic ingestion of high concentrations of ethanol. Journal of Biological Chemistry, *289*(22), 15449-15462.

[26] Zhang, X., Wang, Z., Li, J., Gu, D., Li, S., Shen, C., & Song, Z. (2013). Increased 4-hydroxynonenal formation contributes to obesity-related lipolytic activation in adipocytes. PLoS One, *8*(8), e70663.

[27] D. Demozay, J. C. Mas, S. Rocchi, E. Van Obberghen, FALDH reverses the deleterious action of oxidative stress induced by lipid peroxidation product 4-hydroxynonenal on insulin signaling in 3T3-L1 adipocytes. Diabetes 57, 1216-1226 (2008).

[28] Xu, X. J., Gauthier, M. S., Hess, D. T., Apovian, C. M., Cacicedo, J. M., Gokce, N., ... & Ruderman, N. B. (2012). Insulin sensitive and resistant obesity in humans: AMPK activity, oxidative stress, and depot-specific changes in gene expression in adipose tissue. Journal of lipid research, *53*(4), 792-801.

[29] Kehrer, J. P., & Biswal, S. S. (2000). The molecular effects of acrolein. Toxicological Sciences, *57*(1), 6-15.

[30] Pillon, N. J., Croze, M. L., Vella, R. E., Soulère, L., Lagarde, M., & Soulage, C. O. (2012). The lipid peroxidation by-product 4-hydroxy-2-nonenal (4-HNE) induces insulin resistance in skeletal muscle through both carbonyl and oxidative stress. Endocrinology, *153*(5), 2099-2111.

[31] Kawasaki, N., Asada, R., Saito, A., Kanemoto, S., & Imaizumi, K. (2012). Obesity-induced endoplasmic reticulum stress causes chronic inflammation in adipose tissue. Scientific reports, 2(1), 1-7.

[32] Lee, J. M., Park, S., Lee, D., Ginting, R. P., Lee, M. R., Lee, M. W., & Han, J. (2021). Reduction in endoplasmic reticulum stress activates beige adipocytes differentiation and alleviates high fat diet-induced metabolic phenotypes. *Biochimica et Biophysica Acta (BBA)-Molecular Basis of Disease*, *1867*(5), 166099.

[33] Frohnert, B. I., Long, E. K., Hahn, W. S., & Bernlohr, D. A. (2014). Glutathionylated lipid aldehydes are products of adipocyte oxidative stress and activators of macrophage inflammation. Diabetes, *63*(1), 89-100.

[34] Rahman, M., Steuer, J., Gillgren, P., Végvári, Á., Liu, A., & Frostegård, J. (2019). Malondialdehyde conjugated with albumin induces pro-Inflammatory activation of T cells isolated from human atherosclerotic plaques both directly and via dendritic cell–mediated mechanism. JACC: Basic to Translational Science, *4*(4), 480-494.

[35] Keuper, M. (2019). On the role of macrophages in the control of adipocyte energy metabolism. Endocrine Connections, *8*(6), R105-R121.

[36] Jiang, S. Z., Lu, W., Zong, X. F., Ruan, H. Y., & Liu, Y. (2016). Obesity and hypertension. Experimental and therapeutic medicine, *12*(4), 2395-2399.

[37] Kargi, A. Y., & Iacobellis, G. (2014). Adipose tissue and adrenal glands: novel pathophysiological mechanisms and clinical applications. *International Journal of Endocrinology, 2014*.

[38] Sarzani, R., Guerra, F., Mancinelli, L., Buglioni, A., Franchi, E., & Dessì-Fulgheri, P. (2012). Plasma aldosterone is increased in class 2 and 3 obese essential hypertensive patients despite drug treatment. American Journal of Hypertension, 25, 818-826

[39] Goodfriend, T. L., Ball, D. L., Egan, B. M., Campbell, W. B., & Nithipatikom, K. (2004). Epoxy-keto derivative of linoleic acid stimulates aldosterone secretion. Hypertension, *43*(2), 358-363.

[40] Cannavo, A., Bencivenga, L., Liccardo, D., Elia, A., Marzano, F., Gambino, G., ... & Paolocci, N. (2018). Aldosterone and mineralocorticoid receptor system in cardiovascular physiology and pathophysiology. *Oxidative medicine and cellular longevity*, Article ID 1204598.

[41] Zhong, Z., Kotova, O., Davidescu, A., Ehren, I., Ekberg, K., Jörnvall, H., ... & Chibalin, A. V. (2004). C-peptide stimulates Na+, K+-ATPase via activation of ERK1/2 MAP kinases in human renal tubular cells. Cellular and Molecular Life Sciences CMLS, *61*(21), 2782-2790.

[42] Beltowski, J., Wójcicka, G., Jamroz-Wiśniewska, A., Borkowska, E., & Marciniak, A. (2005). Antioxidant treatment normalizes nitric oxide production, renal sodium handling and blood pressure in experimental hyperleptinemia. Life sciences, *77*(15), 1855-1868.

[43] Beltowski, J. (2010). Leptin and the regulation of renal sodium handling and renal Na+-transporting ATPases: role in the pathogenesis of arterial hypertension. Current cardiology reviews, *6*(1), 31-40.

[44] Packer, M. (2020). Interplay of adenosine monophosphate-activated protein kinase/sirtuin-1 activation and sodium influx inhibition mediates the renal benefits of sodium-glucose co-transporter-2 inhibitors in type 2 diabetes: A novel conceptual framework. Diabetes, Obesity and Metabolism, *22*(5), 734-742.

[45] Asselin, C., Bouchard, B., Tardif, J. C., & Des Rosiers, C. (2006). Circulating 4-hydroxynonenal–protein thioether adducts assessed by gas chromatography–mass spectrometry are increased with disease progression and aging in spontaneously hypertensive rats. Free Radical Biology and Medicine, *41*(1), 97-105.

[46] Aldini, G., Orioli, M., Rossoni, G., Savi, F., Braidotti, P., Vistoli, G., ... & Carini, M. (2011). The carbonyl scavenger carnosine ameliorates dyslipidaemia and renal function in Zucker obese rats. Journal of cellular and molecular medicine, *15*(6), 1339-1354.

[47] Banday, A. A., & Lokhandwala, M. F. (2008). Oxidative stress-induced renal angiotensin AT1 receptor upregulation causes increased stimulation of sodium transporters and hypertension. American Journal of Physiology-Renal Physiology, *295*(3), F698-F706.

[48] Luo, P., Yan, M., Frohlich, E. D., Mehta, J. L., & Hu, C. (2011). Novel concepts in the genesis of hypertension: role of LOX-1. Cardiovascular drugs and therapy, 25(5), 441.

[49] Frostegård, J., Wu, R., Lemne, C., Thulin, T., Witztum, J. L., & de Faire, U. (2003). Circulating oxidized low-density lipoprotein is increased in hypertension. Clinical Science, *105*(5), 615-620.

[50] Jung, S. B., Kim, C. S., Kim, Y. R., Naqvi, A., Yamamori, T., Kumar, S., ... & Irani, K. (2013). Redox factor-1 activates endothelial SIRTUIN1 through reduction of conserved cysteine sulfhydryls in its deacetylase domain. PloS one, *8*(6), e65415.

[51] Xie, X., Yang, C., Cui, Q., Ma, W., Liu, J., Yao, Q., ... & Wang, N. (2019). Stachydrine Mediates Rapid Vascular Relaxation: Activation of Endothelial Nitric Oxide Synthase Involving AMP-Activated Protein Kinase and Akt Phosphorylation in Vascular Endothelial Cells. Journal of Agricultural and Food Chemistry, *67*(35), 9805-9811.

[52] Yang, L., Zhang, J., Xing, W., Zhang, X., Xu, J., Zhang, H., ... & Gao, F. (2016). SIRT3 deficiency induces endothelial insulin resistance and blunts endothelial-dependent vasorelaxation in mice and human with obesity. Scientific reports, *6*(1), 1-12.

[53] Vladykovskaya, E., Sithu, S. D., Haberzettl, P., Wickramasinghe, N. S., Merchant, M. L., Hill, B. G., ... & Srivastava, S. (2012). Lipid peroxidation product 4-hydroxy-trans-2-nonenal causes endothelial activation by inducing endoplasmic reticulum stress. Journal of Biological Chemistry, *287*(14), 11398-11409.

[54] Dehghani, T., & Panitch, A. (2020). Endothelial cells, neutrophils and platelets: getting to the bottom of an inflammatory triangle. Open Biology, *10*(10), 200161.

[55] Shaeib, F., Khan, S. N., Thakur, M., Kohan-Ghadr, H. R., Drewlo, S., Saed, G. M., ... & Abu-Soud, H. M. (2016). The impact of myeloperoxidase and activated macrophages on metaphase II mouse oocyte quality. PLoS One, 11(3), e0151160.

[56] Delporte, C., Van Antwerpen, P., Vanhamme, L., Roumeguère, T., & Zouaoui Boudjeltia, K. (2013). Low-density lipoprotein modified by

myeloperoxidase in inflammatory pathways and clinical studies. *Mediators of inflammation*, *2013*.

[57] Azcona, J. A., Tang, S., Jeitner, T., Guo, A. M., & Schwartzman, M. L. (2019). Novel Contribution of Myeloperoxidase-derived Hypochlorous Acid to 20-HETE Production Post-Ischemia. The FASEB Journal, *33*(S1), 513-511.

[58] Wu, C. C., & Schwartzman, M. L. (2011). The role of 20-HETE in androgen-mediated hypertension. Prostaglandins & other lipid mediators, *96*(1-4), 45-53.

[59] Ecobici, M., & Stoicescu, C. (2017). Arterial Stiffness and Hypertension–Which Comes First?. Maedica, *12*(3), 184.

[60] Soler, A., Hunter, I., Joseph, G., Hutcheson, R., Hutcheson, B., Yang, J., ... & Rocic, P. (2018). Elevated 20-HETE in metabolic syndrome regulates arterial stiffness and systolic hypertension via MMP12 activation. Journal of molecular and cellular cardiology, *117*, 88-99.

[61] Borregaard, N., Jensen, H. S., & Bjerrum, O. W. (1987). Prevention of tissue damage: Inhibition of Myeloperoxidase mediated inactivation of α 1 inhibitor by N-acetyl cysteine, Glutathione, and Methionine. Agents and Actions, *22*(3), 255-260.

[62] van der Toorn, M., Smit-de Vries, M. P., Slebos, D. J., de Bruin, H. G., Abello, N., van Oosterhout, A. J., ... & Kauffman, H. F. (2007). Cigarette smoke irreversibly modifies glutathione in airway epithelial cells. American Journal of Physiology-Lung Cellular and Molecular Physiology, *293*(5), L1156-L1162.

[63] Safar, M. E., Asmar, R., Benetos, A., Blacher, J., Boutouyrie, P., Lacolley, P., ... & French Study Group on Arterial Stiffness. (2018). Interaction between hypertension and arterial stiffness: an expert reappraisal. Hypertension, 72(4), 796-805.

[64] Ren, R., Hashimoto, T., Mizuno, M., Takigawa, H., Yoshida, M., Azuma, T., & Kanazawa, K. (2013). A lipid peroxidation product 9-oxononanoic acid induces phospholipase A2 activity and thromboxane A2 production in human blood. Journal of Clinical Biochemistry and Nutrition, *52*(3), 228-233.

[65] Zhao, W. G., & Richardson, J. S. (1990). Prostacyclin, thromboxane A2, and hypertension. *Clinical and investigative medicine.* Medecine clinique et experimentale, *13*(6), 343-352.

[66] Slatter, D. A., Paul, R. G., Murray, M., & Bailey, A. J. (1999). Reactions of lipid-derived malondialdehyde with collagen. Journal of Biological Chemistry, *274*(28), 19661-19669.

[67] Kawamoto, R., Ninomiyax, D., Kusunoki, T., Kasai, Y., Ohtsuka, N., & Kumagi, T. (2016). Oxidative stress is associated with increased arterial stiffness in middle-aged and elderly community-dwelling persons. Journal of Clinical Gerontology and Geriatrics, *7*(4), 136-140.

[68] Sell, D. R., & Monnier, V. M. (2012). Molecular basis of arterial stiffening: role of glycation–a mini-review. Gerontology, *58*(3), 227-237.

[69] Grossin, N., Auger, F., Niquet-Leridon, C., Durieux, N., Montaigne, D., Schmidt, A. M., ... & Boulanger, E. (2015). Dietary CML-enriched protein induces functional arterial aging in a RAGE-dependent manner in mice. Molecular nutrition & food research, *59*(5), 927-938.

[70] Peng, Y., Kim, J. M., Park, H. S., Yang, A., Islam, C., Lakatta, E. G., & Lin, L. (2016). AGE-RAGE signal generates a specific NF-κB RelA

"barcode" that directs collagen I expression. Scientific reports, *6*(1), 1-10.

[71] Han, J. M., Kim, H. I., Lee, Y. J., Lee, J. W., Kim, K. M., & Bae, J. C. (2019). Differing associations between fatty liver and dyslipidemia according to the degree of hepatic steatosis in Korea. Journal of Lipid and Atherosclerosis 8, 258-266.

[72] Chen, Z., Tian, R., She, Z., Cai, J., & Li, H. (2020). Role of oxidative stress in the pathogenesis of nonalcoholic fatty liver disease. Free Radical Biology and Medicine, *152*, 116-141.

[73] Chen, L., Zhu, H., Su, S., Harshfield, G., Sullivan, J., Webb, C., ... & Dong, Y. (2020). High-mobility group box-1 is associated with obesity, inflammation, and subclinical cardiovascular risk among young adults: a longitudinal cohort study. Arteriosclerosis, thrombosis, and vascular biology, *40*(11), 2776-2784.

[74] Morita, M., Ishida, N., Uchiyama, K., Yamaguchi, K., Itoh, Y., Shichiri, M., ... & Niki, E. (2012). Fatty liver induced by free radicals and lipid peroxidation. Free radical research, *46*(6), 758-765.

[75] Yu, S. S., Castillo, D. C., Courville, A. B., & Sumner, A. E. (2012). The triglyceride paradox in people of African descent. Metabolic syndrome and related disorders, *10*(2), 77-82.

[76] Kovacova, Z., Tencerova, M., Roussel, B., Wedellova, Z., Rossmeislova, L., Langin, D., ... & Stich, V. (2012). The impact of obesity on secretion of adiponectin multimeric isoforms differs in visceral and subcutaneous adipose tissue. International Journal of Obesity, *36*(10), 1360-1365.

[77] Clarenbach, J. J., Vega, G. L., Adams-Huet, B., Considine, R. V., Ricks, M., & Sumner, A. E. (2007). Variability in postheparin hepatic lipase activity is associated with plasma adiponectin levels in African Americans. Journal of Investigative Medicine, *55*(4), 187-194.

[78] Marfella, R., Grella, R., Rizzo, M. R., Barbieri, M., Grella, R., Ferraraccio, F., ... & Nicoletti, G. (2009). Role of subcutaneous abdominal fat on cardiac function and proinflammatory cytokines in premenopausal obese women. Annals of plastic surgery, *63*(5), 490-495.

[79] Haas, M. J., & Mooradian, A. D. (2010). Regulation of high-density lipoprotein by inflammatory cytokines: establishing links between immune dysfunction and cardiovascular disease. Diabetes/metabolism research and reviews, 26(2), 90-99.

[80] Singh, R., Wang, Y., Schattenberg, J. M., Xiang, Y., & Czaja, M. J. (2009). Chronic oxidative stress sensitizes hepatocytes to death from 4-hydroxynonenal by JNK/c-Jun overactivation. American Journal of Physiology-Gastrointestinal and Liver Physiology, *297*(5), G907-G917.

[81] Onyango, A. N. (2020). Mechanisms of the regulation and dysregulation of glucagon secretion. *Oxidative Medicine and Cellular Longevity*, Article ID 3089139

[82] Rabinovitch, A. L. E. X., Suarez-Pinzon, W. L., Strynadka, K., Lakey, J. R., & Rajotte, R. V. (1996). Human pancreatic islet beta-cell destruction by cytokines involves oxygen free radicals and aldehyde production. The Journal of Clinical Endocrinology & Metabolism, *81*(9), 3197-3202.

[83] Femlak, M., Gluba-Brzózka, A., Ciałkowska-Rysz, A., & Rysz, J. (2017). The role and function of HDL in patients with diabetes mellitus and the related cardiovascular risk. Lipids in Health and Disease, *16*(1), 1-9.

[84] Shao, B., Pennathur, S., Pagani, I., Oda, M. N., Witztum, J. L., Oram, J. F., & Heinecke, J. W. (2010). Modifying apolipoprotein AI by malondialdehyde, but not by an array of other reactive carbonyls, blocks cholesterol efflux by the ABCA1 pathway. Journal of Biological Chemistry, 285(24), 18473-18484.

[85] Huang, J., Yancey, P. G., Tao, H., Borja, M. S., Smith, L. E., Kon, V., ... & Linton, M. F. (2020). Reactive dicarbonyl scavenging effectively reduces MPO-mediated oxidation of HDL and restores PON1 activity. Nutrients, *12*(7), 1937.

[86] Plaisance, V., Brajkovic, S., Tenenbaum, M., Favre, D., Ezanno, H., Bonnefond, A., ... & Abderrahmani, A. (2016). Endoplasmic reticulum stress links oxidative stress to impaired pancreatic beta-cell function caused by human oxidized LDL. PLoS One, *11*(9), e0163046.

[87] Mazier, W., & Cota, D. (2017). Islet endothelial cell: friend and foe. Endocrinology, *158*(2), 226-228.

[88] Kowluru, A. (2020). Oxidative Stress in Cytokine-Induced Dysfunction of the Pancreatic Beta Cell: Known Knowns and Known Unknowns. Metabolites, *10*(12), 480.

[89] Tran, T. T. T., Postal, B. G., Demignot, S., Ribeiro, A., Osinski, C., de Barros, J. P. P., ... & Carrière, V. (2016). Short term palmitate supply impairs intestinal insulin signaling via ceramide production. Journal of Biological Chemistry, *291*(31), 16328-16338.

[90] Fucho, R., Casals, N., Serra, D., & Herrero, L. (2017). Ceramides and mitochondrial fatty acid oxidation in obesity. The FASEB Journal, 31(4), 1263-1272.

[91] Haffar, T., Akoumi, A., & Bousette, N. (2016). Lipotoxic palmitate impairs the rate of β-oxidation and citric acid cycle flux in rat neonatal cardiomyocytes. Cellular Physiology and Biochemistry, *40*(5), 969-981.

[92] Wang, J., Chen, Y., Song, Q., Griffiths, A., & Song, Z. (2020). mTORC1-IRE1α pathway activation contributes to palmitate-elicited triglyceride secretion and cell death in hepatocytes. Experimental Biology and Medicine, *245*(14), 1268-1279.

[93] Tang, Y., Zhao, X., Antoine, D., Xiao, X., Wang, H., Andersson, U., ... & Lu, B. (2016). Regulation of posttranslational modifications of HMGB1 during immune responses. Antioxidants & Redox Signaling, *24*(12), 620-634.

[94] Nair, A., Ebenezer, P., & Francis, J. (2015). High-Mobility Group Box 1 (HMGB1) in Angiotensin II induced hypertension. The FASEB Journal, *29*, 1059-1057.

[95] Wang, Y., Zhong, J., Zhang, X., Liu, Z., Yang, Y., Gong, Q., & Ren, B. (2016). The role of HMGB1 in the pathogenesis of type 2 diabetes. Journal of diabetes research, *2016*.

[96] Schüler, R., Osterhoff, M. A., Frahnow, T., Seltmann, A. C., Busjahn, A., Kabisch, S., ... & Pfeiffer, A. F. (2017). High-saturated-fat diet increases circulating angiotensin-converting enzyme, which is enhanced by the rs4343 polymorphism defining persons at risk of nutrient-dependent increases of blood pressure. Journal of the American Heart Association, *6*(1), e004465.

[97] Nieva-Echevarría, B., Goicoechea, E., & Guillén, M. D. (2020). Food lipid oxidation under gastrointestinal digestion conditions: A review. Critical Reviews in Food Science and Nutrition, *60*(3), 461-478

[98] Macho-González, A., Garcimartín, A., López-Oliva, M. E., Bastida, S.,

Benedí, J., Ros, G., ... & Sánchez-Muniz, F. J. (2020). Can Meat and Meat-Products Induce Oxidative Stress?. Antioxidants, *9*(7), 638.

[99] Kanner, J., Gorelik, S., Roman, S., & Kohen, R. (2012). Protection by polyphenols of postprandial human plasma and low-density lipoprotein modification: the stomach as a bioreactor. Journal of Agricultural and Food Chemistry, *60*(36), 8790-8796.

[100] Cheon, H. G., & Cho, Y. S. (2014). Protection of palmitic acid-mediated lipotoxicity by arachidonic acid via channeling of palmitic acid into triglycerides in C2C12. Journal of biomedical science, *21*(1), 1-10.

[101] Moravcova, A., Červinková, Z., Kučera, O., Mezera, V., Rychtrmoc, D., & Lotkova, H. (2015). The effect of oleic and palmitic acid on induction of steatosis and cytotoxicity on rat hepatocytes in primary culture. Physiological research, *64*.

[102] Zhang, Y., Xue, R., Zhang, Z., Yang, X., & Shi, H. (2012). Palmitic and linoleic acids induce ER stress and apoptosis in hepatoma cells. Lipids in health and disease, *11*(1), 1-8.

[103] Mak, I. L., Lavery, P., Agellon, S., Rauch, F., Murshed, M., & Weiler, H. A. (2019). Arachidonic acid exacerbates diet-induced obesity and reduces bone mineral content without impacting bone strength in growing male rats. The Journal of nutritional biochemistry, *73*, 108226.

[104] Fleckenstein-Elsen, M., Dinnies, D., Jelenik, T., Roden, M., Romacho, T., & Eckel, J. (2016). Eicosapentaenoic acid and arachidonic acid differentially regulate adipogenesis, acquisition of a brite phenotype and mitochondrial function in primary human adipocytes. Molecular nutrition & food research, *60*(9), 2065-2075.

[105] Pisani, D. F., Ghandour, R. A., Beranger, G. E., Le Faouder, P., Chambard, J. C., Giroud, M., ... & Amri, E. Z. (2014). The ω6-fatty acid, arachidonic acid, regulates the conversion of white to brite adipocyte through a prostaglandin/calcium mediated pathway. Molecular metabolism, *3*(9), 834-847.

[106] Chan, P. C., Hsiao, F. C., Chang, H. M., Wabitsch, M., & Hsieh, P. S. (2016). Importance of adipocyte cyclooxygenase-2 and prostaglandin E2-prostaglandin E receptor 3 signaling in the development of obesity-induced adipose tissue inflammation and insulin resistance. The FASEB Journal, *30*(6), 2282-2297.

[107] Uchida, K. (2017). HNE as an inducer of COX-2. Free Radical Biology and Medicine, *111*, 169-172.

[108] Williams, E. S., Baylin, A., & Campos, H. (2007). Adipose tissue arachidonic acid and the metabolic syndrome in Costa Rican adults. Clinical nutrition, *26*(4), 474-482.

[109] Gilani, A., Agostinucci, K., Hossain, S., Pascale, J. V., Garcia, V., Adebesin, A. M., ... & Schwartzman, M. L. (2021). 20-HETE interferes with insulin signaling and contributes to obesity-driven insulin resistance. Prostaglandins & Other Lipid Mediators, *152*, 106485.

[110] Nelson, G. J., Schmidt, P. C., Bartolini, G., Kelley, D. S., Phinney, S. D., Kyle, D., ... & Schaefer, E. J. (1997). The effect of dietary arachidonic acid on plasma lipoprotein distributions, apoproteins, blood lipid levels, and tissue fatty acid composition in humans. Lipids, *32*(4), 427-433.

[111] Froyen, E., & Burns-Whitmore, B. (2020). The Effects of Linoleic Acid Consumption on Lipid Risk Markers for Cardiovascular Disease in Healthy Individuals: A Review of Human

Intervention Trials. Nutrients, *12*(8), 2329.

[112] Iggman, D., Ärnlöv, J., Cederholm, T., & Risérus, U. (2016). Association of adipose tissue fatty acids with cardiovascular and all-cause mortality in elderly men. JAMA cardiology, *1*(7), 745-753.

[113] Jandacek, R. J. (2017, June). Linoleic acid: a nutritional quandary. In *Healthcare* (Vol. 5, No. 2, p. 25). Multidisciplinary Digital Publishing Institute.

[114] Kothapalli, K. S., Ye, K., Gadgil, M. S., Carlson, S. E., O'Brien, K. O., Zhang, J. Y., ... & Brenna, J. T. (2016). Positive selection on a regulatory insertion–deletion polymorphism in FADS2 influences apparent endogenous synthesis of arachidonic acid. Molecular Biology and Evolution, *33*(7), 1726-1739.

[115] Li, P., Zhao, J., Kothapalli, K. S., Li, X., Li, H., Han, Y., ... & Gao, Y. (2018). A regulatory insertion-deletion polymorphism in the FADS gene cluster influences PUFA and lipid profiles among Chinese adults: a population-based study. The American Journal of Clinical Nutrition, *107*(6), 867-875.

[116] Goedecke, J. H., Chorell, E., van Jaarsveld, P. J., Risérus, U., & Olsson, T. (2021). Fatty acid metabolism and associations with insulin sensitivity differs between black and white South African women. The Journal of Clinical Endocrinology & Metabolism, *106*(1), e140-e151.

[117] Tsukamoto, I., & Sugawara, S. (2018). Low levels of linoleic acid and α-linolenic acid and high levels of arachidonic acid in plasma phospholipids are associated with hypertension. Biomedical reports, *8*(1), 69-76.

[118] Maingrette, F., & Renier, G. (2005). Linoleic acid increases lectin-like oxidized LDL receptor-1 (LOX-1) expression in human aortic endothelial cells. Diabetes, *54*(5), 1506-1513.

[119] Hao, Q., Lillefosse, H. H., Fjære, E., Myrmel, L. S., Midtbø, L. K., Jarlsby, R. H., ... & Madsen, L. (2012). High-glycemic index carbohydrates abrogate the antiobesity effect of fish oil in mice. American Journal of Physiology-Endocrinology and Metabolism, *302*(9), E1097-E1112.

[120] Zou, Y., Wang, Y. N., Ma, H., He, Z. H., Tang, Y., Guo, L., ... & Tang, Q. Q. (2020). SCD1 promotes lipid mobilization in subcutaneous white adipose tissue. Journal of lipid research, *61*(12), 1589-1604.

[121] Camell, C., & Smith, C. W. (2013). Dietary oleic acid increases M2 macrophages in the mesenteric adipose tissue. PloS one, *8*(9), e75147.

[122] Guo, X. F., Li, X., Shi, M., & Li, D. (2017). n-3 polyunsaturated fatty acids and metabolic syndrome risk: a meta-analysis. Nutrients, 9(7), 703.

[123] Kim, J., Okla, M., Erickson, A., Carr, T., Natarajan, S. K., & Chung, S. (2016). Eicosapentaenoic acid potentiates brown thermogenesis through FFAR4-dependent up-regulation of miR-30b and miR-378. Journal of Biological Chemistry, *291*(39), 20551-20562.

[124] Holland, W. L., Adams, A. C., Brozinick, J. T., Bui, H. H., Miyauchi, Y., Kusminski, C. M., ... & Scherer, P. E. (2013). An FGF21-adiponectin-ceramide axis controls energy expenditure and insulin action in mice. Cell metabolism, *17*(5), 790-797.

[125] Dutchak, P. A., Katafuchi, T., Bookout, A. L., Choi, J. H., Ruth, T. Y., Mangelsdorf, D. J., & Kliewer, S. A. (2012). Fibroblast growth factor-21 regulates PPARγ activity and the

antidiabetic actions of thiazolidinediones. Cell, *148*(3), 556-567.

[126] Quesada-López, T., Cereijo, R., Turatsinze, J. V., Planavila, A., Cairó, M., Gavaldà-Navarro, A., ... & Villarroya, F. (2016). The lipid sensor GPR120 promotes brown fat activation and FGF21 release from adipocytes. Nature communications, *7*(1), 1-17.

[127] Drouin, G., Catheline, D., Guillocheau, E., Gueret, P., Baudry, C., Le Ruyet, P., ... & Legrand, P. (2019). Comparative effects of dietary n-3 docosapentaenoic acid (DPA), DHA and EPA on plasma lipid parameters, oxidative status and fatty acid tissue composition. The Journal of nutritional biochemistry, *63*, 186-196.

[128] Griffin, B. A. (2008). How relevant is the ratio of dietary n-6 to n-3 polyunsaturated fatty acids to cardiovascular disease risk? Evidence from the OPTILIP study. Current opinion in lipidology, *19*(1), 57-62.

[129] Li, N., Jia, M., Deng, Q., Wang, Z., Huang, F., Hou, H., & Xu, T. (2020). Effect of low-ratio n-6/n-3 PUFA on blood lipid level: a meta-analysis. Hormones, 1-10.

[130] Ma, T., Liaset, B., Hao, Q., Petersen, R. K., Fjære, E., Ngo, H. T., ... & Madsen, L. (2011). Sucrose counteracts the anti-inflammatory effect of fish oil in adipose tissue and increases obesity development in mice. PloS one, *6*(6), e21647.

[131] Varma, V., Boros, L. G., Nolen, G. T., Chang, C. W., Wabitsch, M., Beger, R. D., & Kaput, J. (2015). Metabolic fate of fructose in human adipocytes: a targeted 13 C tracer fate association study. Metabolomics, 11(3), 529-544.

[132] Sodhi, K., Hilgefort, J., Banks, G., Gilliam, C., Stevens, S., Ansinelli, H. A., ... & Khitan, Z. (2016). Uric acid-induced adipocyte dysfunction is attenuated by HO-1 upregulation: potential role of antioxidant therapy to target obesity. Stem cells international, 8197325 .

[133] Sautin, Y. Y., & Johnson, R. J. (2008). Uric acid: the oxidant-antioxidant paradox. Nucleosides, Nucleotides, and Nucleic Acids, 27(6-7), 608-619.

[134] Zhang, J., Diao, B., Lin, X., Xu, J., & Tang, F. (2019). TLR2 and TLR4 mediate an activation of adipose tissue renin-angiotensin system induced by uric acid. *Biochimie*, *162*, 125-133.

[135] Bartlett, D. E., Miller, R. B., Thiesfeldt, S., Lakhani, H. V., Shapiro, J. I., & Sodhi, K. (2018). The role of Na/K-ATPase signaling in oxidative stress related to aging: implications in obesity and cardiovascular disease. *International journal of molecular sciences*, *19*(7), 2139.

[136] Pratt, R. D., Brickman, C., Nawab, A., Cottrill, C., Snoad, B., Lakhani, H. V., ... & Sodhi, K. (2019). The adipocyte Na/K-ATPase oxidant amplification loop is the central regulator of western diet-induced obesity and associated comorbidities. *Scientific reports*, *9*(1), 1-11.

[137] Lee, M., Sorn, S. R., Lee, Y., & Kang, I. (2019). Salt induces adipogenesis/lipogenesis and inflammatory adipocytokines secretion in adipocytes. International journal of molecular sciences, 20(1), 160.

[138] Diao, Z., Asico, L. D., Van Anthony, M. V., Zheng, X., Cuevas, S., Armando, I., ... & Wang, X. (2017). Increased renal oxidative stress in salt-sensitive human GRK4γ486V transgenic mice. Free Radical Biology and Medicine, 106, 80-90.

[139] Zhu, Z., Wu, F., Lu, Y., Wang, Z., Zang, J., Yu, H., ... & Ding, G. (2018).

The association of dietary cholesterol and fatty acids with dyslipidemia in Chinese metropolitan men and women. Nutrients, 10(8), 961

[140] Püschel, G. P., & Henkel, J. (2018). Dietary cholesterol does not break your heart but kills your liver. Porto Biomedical Journal, 3, e12.

[141] Sánchez-Tapia, M., Miller, A. W., Granados-Portillo, O., Tovar, A. R., & Torres, N. (2020). The development of metabolic endotoxemia is dependent on the type of sweetener and the presence of saturated fat in the diet. Gut Microbes, 12, 1801301.

[142] Bonet, M. L., Canas, J. A., Ribot, J., & Palou, A. (2015). Carotenoids and their conversion products in the control of adipocyte function, adiposity and obesity. Archives of biochemistry and biophysics, 572, 112-125.

[143] Wong, S. K., Chin, K. Y., Suhaimi, F. H., Ahmad, F., & Ima-Nirwana, S. (2017). Vitamin E as a potential interventional treatment for metabolic syndrome: evidence from animal and human studies. Frontiers in pharmacology, 8, 444.

[144] Puah, C., Choo, Y., Ma, A., & Chuah, C. (2007). The effect of physical refining on palm vitamin E (tocopherol, tocotrienol and tocomonoenol). American Journal of Applied Sciences, 4(6), 374-377.

[145] Sotler, R., Poljšak, B., Dahmane, R., Jukić, T., Jukić, D. P., Rotim, C., ... & Starc, A. (2019). Prooxidant activities of antioxidants and their impact on health. *Acta Clinica Croatica*, *58*(4), 726.

[146] Zou, W., Chen, C., Zhong, Y., An, J., Zhang, X., Yu, Y., ... & Fu, J. (2013). PI3K/Akt pathway mediates Nrf2/ARE activation in human L02 hepatocytes exposed to low-concentration HBCDs. Environmental science & technology, 47(21), 12434-12440.

[147] Wardyn, J. D., Ponsford, A. H., & Sanderson, C. M. (2015). Dissecting molecular cross-talk between Nrf2 and NF-κB response pathways. Biochemical society transactions, 43(4), 621-626.

[148] Bonet, M. L., Canas, J. A., Ribot, J., & Palou, A. (2015). Carotenoids and their conversion products in the control of adipocyte function, adiposity and obesity. Archives of biochemistry and biophysics, 572, 112-125.

[149] Wong, S. K., Chin, K. Y., Suhaimi, F. H., Ahmad, F., & Ima-Nirwana, S. (2017). Vitamin E as a potential interventional treatment for metabolic syndrome: evidence from animal and human studies. Frontiers in pharmacology, 8, 444.

[150] Puah, C., Choo, Y., Ma, A., & Chuah, C. (2007). The effect of physical refining on palm vitamin E (tocopherol, tocotrienol and tocomonoenol). American Journal of Applied Sciences, 4(6), 374-377.

[151] Zhuang, Y., Wu, H., Wang, X., He, J., He, S., & Yin, Y. (2019). Resveratrol attenuates oxidative stress-induced intestinal barrier injury through PI3K/Akt-mediated Nrf2 signaling pathway. Oxidative medicine and cellular longevity, 2019.

[152] He, S., Guo, Y., Zhao, J., Xu, X., Wang, N., & Liu, Q. (2020). Ferulic acid ameliorates lipopolysaccharide-induced barrier dysfunction via microRNA200c-3p-mediated activation of PI3K/AKT pathway in Caco-2 cells. Frontiers in pharmacology, 11, 376.

[153] Zhao, L., Zhang, Q., Ma, W., Tian, F., Shen, H., & Zhou, M. (2017). A combination of quercetin and resveratrol reduces obesity in high-fat diet-fed rats by modulation of gut microbiota. Food & function, 8(12), 4644-4656.

[154] Zhao, Y., Chen, B., Shen, J., Wan, L., Zhu, Y., Yi, T., & Xiao, Z. (2017).

The beneficial effects of quercetin, curcumin, and resveratrol in obesity. Oxidative medicine and cellular longevity, 2017.

[155] Xu, T., Song, Q., Zhou, L., Yang, W., Wu, X., Qian, Q., ... & Li, S. (2021). Ferulic acid alleviates lipotoxicityinduced hepatocellular death through the SIRT1-regulated autophagy pathway and independently of AMPK and Akt in AML-12 hepatocytes. Nutrition & metabolism, 18(1), 1-13.

[156] Ruan, H., & Dong, L. Q. (2016). Adiponectin signaling and function in insulin target tissues. Journal of molecular cell biology, 8(2), 101-109.

[157] Miyata, Y., Tanaka, H., Shimada, A., Sato, T., Ito, A., Yamanouchi, T., & Kosano, H. (2011). Regulation of adipocytokine secretion and adipocyte hypertrophy by polymethoxyflavonoids, nobiletin and tangeretin. Life sciences, 88(13-14), 613-618.

[158] Rezvan, N., Moini, A., GorganiFiruzjaee, S., & Hosseinzadeh-Attar, M. J. (2018). Oral quercetin supplementation enhances adiponectin receptor transcript expression in polycystic ovary syndrome patients: a randomized placebo-controlled doubleblind clinical trial. Cell Journal (Yakhteh), 19(4), 627.

[159] Shahi, M., Zakerzadeh, M., Zakerkish, M., Zarei, M., & Saki, A. (2017). Effect of sesamin supplementation on glycemic status, inflammatory markers, and adiponectin levels in patients with type 2 diabetes mellitus. Journal of dietary supplements, 14(1), 65-75.

[160] Tomé-Carneiro, J., Gonzálvez, M., Larrosa, M., Yáñez-Gascón, M. J., García-Almagro, F. J., Ruiz-Ros, J. A., ... & Espín, J. C. (2013). Grape resveratrol increases serum adiponectin and downregulates inflammatory genes in peripheral blood mononuclear cells: a triple-blind, placebo-controlled, one-year clinical trial in patients with stable coronary artery disease. Cardiovascular Drugs and Therapy, 27(1), 37-48.

[161] Reibe-Pal, S., & Febbraio, M. A. (2017). Adiponectin serenades ceramidase to improve metabolism. Molecular metabolism, 6(3), 233.

[162] Jin, J., Lu, Z., Li, Y., Cowart, L. A., Lopes-Virella, M. F., & Huang, Y. (2018). Docosahexaenoic acid antagonizes the boosting effect of palmitic acid on LPS inflammatory signaling by inhibiting gene transcription and ceramide synthesis. PloS one, 13(2), e0193343.

[163] Deng, P., Hoffman, J. B., Petriello, M. C., Wang, C. Y., Li, X. S., Kraemer, M. P., ... & Hennig, B. (2020). Dietary inulin decreases circulating ceramides by suppressing neutral sphingomyelinase expression and activity in mice. Journal of lipid research, 61(1), 45-53.

[164] Mihalj, M., Tadzic, R., Vcev, A., Rucevic, S., & Drenjancevic, I. (2016). Blood pressure reduction is associated with the changes in oxidative stress and endothelial activation in hypertension, regardless of antihypertensive therapy. Kidney and Blood Pressure Research, 41(6), 721-735.

[165] Rao, P. V., & Gan, S. H. (2014). Cinnamon: a multifaceted medicinal plant. Evidence-Based Complementary and Alternative Medicine, 642942.

[166] Lundberg, J. O., Carlström, M., & Weitzberg, E. (2018). Metabolic effects of dietary nitrate in health and disease. Cell Metabolism, 28(1), 9-22.

[167] Bunbupha, S., Prasarttong, P., Poasakate, A., Maneesai, P., & Pakdeechote, P. (2021). Imperatorin alleviates metabolic and vascular alterations in high-fat/high-fructose

diet-fed rats by modulating adiponectin receptor 1, eNOS, and p47phox expression. European Journal of Pharmacology, 174010.

[168] Anderson, E. J., Vistoli, G., Katunga, L. A., Funai, K., Regazzoni, L., Monroe, T. B., ... & Aldini, G. (2018). A carnosine analog mitigates metabolic disorders of obesity by reducing carbonyl stress. The Journal of clinical investigation, 128(12), 5280-5293.

[169] Ho, C. T., & Wang, M. (2013). Dietary phenolics as reactive carbonyl scavengers: potential impact on human health and mechanism of action. Journal of traditional and complementary medicine, 3(3), 139-141.

[170] Khalifa, I., Xia, D., Dutta, K., Peng, J., Jia, Y., & Li, C. (2020). Mulberry anthocyanins exert anti-AGEs effects by selectively trapping glyoxal and structural-dependently blocking the lysyl residues of β-lactoglobulins. Bioorganic Chemistry, 96, 103615.

[171] Wang, X., Chang, T., Jiang, B., Desai, K., & Wu, L. (2007). Attenuation of hypertension development by aminoguanidine in spontaneously hypertensive rats: role of methylglyoxal. American journal of hypertension, 20(6), 629-636.

[172] Liu, P., Wang, Y., Yang, G., Zhang, Q., Meng, L., Xin, Y., & Jiang, X. (2021). The role of short-chain fatty acids in intestinal barrier function, inflammation, oxidative stress, and colonic carcinogenesis. Pharmacological Research, 105420.

[173] Yan, H., & Ajuwon, K. M. (2017). Butyrate modifies intestinal barrier function in IPEC-J2 cells through a selective upregulation of tight junction proteins and activation of the Akt signaling pathway. PloS one, 12(6), e0179586.

[174] Mani, V., Hollis, J. H., & Gabler, N. K. (2013). Dietary oil composition differentially modulates intestinal endotoxin transport and postprandial endotoxemia. Nutrition & metabolism, 10(1), 1-9.

Chapter 5

Phytotherapeutics Attenuation of Oxidative Stress, Inflammation and Lipid Peroxidation in Severe and Chronic Diseases

Alfred Mavondo-Nyajena Mukuwa Greanious, Nesisa Ncube, Alfred Sibanda, Delton Dube, Francis Chikuse Farai and Paul Makoni

Abstract

Lipid peroxidation is an end process of cellular injury driven by oxidative stress (OS) and inflammation through several molecular changes. Metabolism-generated reactive oxygen species avidly attack the polyunsaturated fatty acids in lipid cell membranes, initiating a self-propagating chain-reaction. Cell membrane destruction, lipids and the end-products of lipid peroxidation reactions are hostile to the viability of cells, even tissues causing and exacerbating Diabetes Mellitus (DM), neurodegenerative disorders (NDDs), cardiovascular diseases (CVDs) and Rheumatoid Arthritis (RA). Current treatment regimens have untoward side effects in the long-term necessitating phytochemical use as these are part of natural food sources. Enzymatic and non-enzymatic antioxidant defense mechanisms may be over run causing lipid peroxidation to take place. In disease states, oxidative stress may increase with subsequent production of increased free radicals which may over run the antioxidant capacity of the body with resultant oxidative damage on polyunsaturated fatty acids in the cell fluid membranes with cellular and tissue damage. Phytochemicals, have been shown to ameliorate diseases through attenuation of oxidative stress, inflammation, lipid peroxidation, causing tissue regeneration by regulating signaling systems and neuroprotective processes. Involvement of polyphenolic and non-phenolic phytochemical in the attenuation of OS, inflammation and lipid peroxidation remain areas of critical importance in combating DM, CVDA, NDD and RA.

Keywords: phytotherapeutics, oxidative stress, inflammation, lipid peroxidation, severe and chronic diseases, phytochemicals

1. Introduction

There is a significant contributory role the Fenton and Haber Weiss reaction makes in oxidative stress (OS) building up to several progressive diseases such as Alzheimer's disease (AD). Reactive oxygen species (ROS) and reactive nitrogen species (RNS) are manufactured by these reactions causing OS in AD. Iron, copper and aluminum influences creation of free radicals such as hydroxyl radicals with impairment to DNA, proteins, lipids and carbohydrates.

Beta amyloid (Aβ42) toxicity is created by hydroxyl (OH^-) radicals from the Fenton reaction [1, 2] in AD. Soluble human fibrinogen is converted into an insoluble fibrin-like aggregate seen in neurodegenerative diseases such as AD when it reacts with hydroxyl radical [3]. The Fenton gated OS attenuates DNA base substitutions of guanine to cytosine when catalyzed by iron and guanine to thymine and cytosine to thymine when catalyzed by copper and Nickel [4].

Phosphoinositide 3- kinase (PI3K), c-Jun N-terminal kinase (JNK), extracellular signal-regulated kinase 1 and 2 (ERK1/2), p38 and transcription factors such as activator protein-1(AP-1), and p53 are signal transduction molecules that are stimulated by ROS [5]. Hastening of AD development are OH^- radicals that impair DNA through p53 pathway. When tumor suppressor gene (TP53) has a mutation, there is increased potential for AD pathogenesis development [6].

The Fenton reaction triggers OS through subtraction of one electron from the molecular oxygen (O_2) resulting in the formation of superoxide (O_2^-) which often produces other ROS species such as H_2O_2 and peroxynitrite $(ONOO)^-$and hydroxyl radicals (OH^-) [7]. This may imply that phytotherapeutics which are able to quench the electron abstraction may alleviate oxidative stress. Moreover, under normal conditions, O_2– has emerged as an important signaling molecule, which regulates precise biochemical reactions and metabolic progressions [8].

The linkage between O_2^-production and H_2O_2 may involve a reduced flavin enzyme by transferring an electron to activate molecular oxygen into superoxide which is either released or enzymatically converted into H_2O_2 [9, 10] or drugs like statins may modify the process [11, 12].

Extreme challenges in the field of ROS-gated diseases is to bridge the knowledge gap between atomic, cellular level and how natural phytochemicals may be used to attenuate OS in certain diseases [13, 14].

The understanding of the Fenton and Haber Weiss reaction and how it may be modified or detoxified or neutralized by phytotherapeutics or nutraceuticals may assist to bridging the knowledge and practice of treating chronic diseases of old age. When phytochemicals stop the subtraction of one electron in Fenton reaction, creation of OS may be averted.

2. Lipid peroxidation products

Characteristics of various lipid peroxidation products as biomarkers have been reviewed on the basis of mechanisms and dynamics of their formation and metabolism and also on the methods of measurement, with an emphasis on the advantages and limitations [15].

Lipid peroxidation or unsaturated lipid reaction with molecular or ROS produces a wide variety of oxidation products with the main primary products being lipid hydroperoxides (LOOH). Many different aldehydes which can be formed as subsequent products during lipid peroxidation include malondialdehyde (MDA), propanal, hexanal, and 4-hydroxynonenal (4-HNE) (**Figure 1**) [16–20].

MDA gives the impression to be the most mutagenic product of lipid peroxidation, whereas 4-HNE is the most toxic [21]. MDA has been extensively used for many years as an expedient biomarker for lipid peroxidation of omega-3 and omega-6 fatty acids because of its simplistic reaction with thiobarbituric acid (TBA) [22]. The TBA test is predicated on the reactivity of TBA toward MDA to yield an intensely colored chromogen fluorescent red adducts. Food chemists used this test initially to evaluate autoxidative degradation of fats and oils [23].

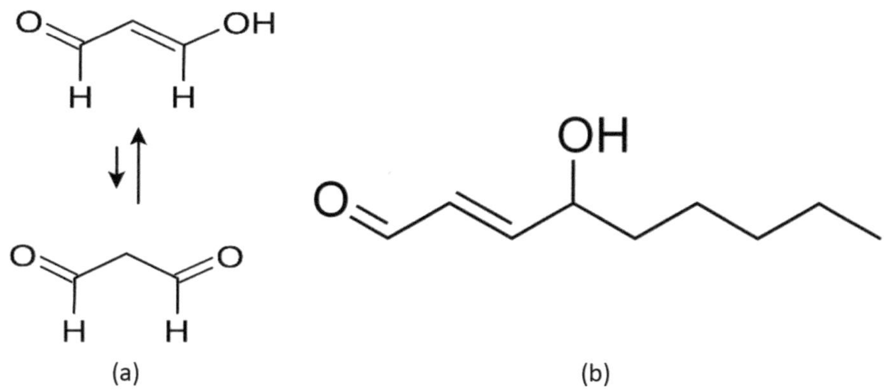

Figure 1.
(a): Structure of 4-hydroxymalondialdehyde and malondialdehyde at equilibrium and (b) 4-hydroxynonenal.

MDA is one of the most popular and reliable biomarkers that determine OS in clinical situations and due to MDA's high reactivity and the toxicity underlying the molecular effect, this molecule is very relevant to biomedical research community [24].

First to be discovered in the 1960s was 4-HNE [25]. Later, in 1980s 4-HNE was described as a cytotoxic product originating from the peroxidation of liver microsomal lipids [26]. The genotoxic effects exerted on human beings results from the subsequently produced 4-HNE from progression of bio membranes lipids peroxidation elicited by free radicals or chemicals [27]. Comparatively large amounts of 4-HNE are produced and they are very reactive aldehydes that act as second messengers of free radicals making them of high significance in disease of old age [28]. Therefore, 4-HNE the most likely easier target for phytotherapeutics as antioxidant in lipid peroxidation.

Also, 4-HNE is a major bioactive marker of lipid peroxidation and a signaling molecule elaborate in regulation of several transcription mediators that are sensitive to stress such as nuclear factor erythroid 2-related factor 2 (Nrf2), activating protein-1 (AP-1), NF-κB, and peroxisome-proliferator-activated receptors (PPAR). These play a critical role in cell proliferation and/or differentiation, cell survival, autophagy, senescence, apoptosis, and necrosis [29]. 4-HNE may stimulate intrinsic and extrinsic apoptotic pathways and interact with typical actors such as tumor protein 53, JNK, Fas or mitochondrial regulators, due to its oxidant status. Simultaneous 4-HNE induces cellular defense mechanisms against OS, thus being involved in its own detoxification and in turn limiting its apoptotic potential [30]. These dualities can imbalance cell fate, either toward cell death or toward survival, depending on the cell type, the metabolic state and the ability to detoxify [31]. The pleiotropism displayed by phytotherapeutics like Asiatic acid [13, 32–36], maslinic acid [37, 38] and oleanolic acid [36] and their involvement in redox reactions may influence 4-HNE activity thus modulating the its function in stress induces pathology of old age.

3. Phytotherapeutics and lipid peroxidation products

Phytotherapeutics which may restore or facilitate the restoration of 4-HNE's apoptotic inhibition potential as well as catalyze its oxidant signaling pathways through direct redox reactions or attenuation of enzymatic antioxidant systems, have a great capacity in modulating lipid peroxidation related diseases [13, 14, 35]. Triterpenes with pleiotropic activities have been shown to possess antioxidant properties as well oxidative functions in certain parasitic infections ameliorating disease outcomes and outputs [33, 39].

Other phytochemicals also follow in this narrative in their use as antioxidants in lipid peroxidation agents and their use to fight against associated diseases using various mechanism. By inhibiting formation of both primary and secondary products of the lipid peroxidation process, plant phytochemicals may exert their effect on hydroperoxide groups from attaching to free fatty acids, triacylglycerols, phospholipids, and sterols [40].

While, hydroperoxides may decompose *in vivo* through two-electron reduction, which may inhibit the peroxidative damage, phytochemicals may also facilitate this process of antioxidant lipid peroxidation through enhancing activity of enzymes for two-electron reduction of hydroperoxides such as selenium-dependent glutathione peroxidases (GPx) and selenoprotein P (SeP) [41, 42].

4. Phytochemical and Antioxidative activity in lipid peroxidation

4.1 *Salix aegyptiaca* and lipid peroxidation

Salix aegyptiaca is a deciduous plant belonging to *Salicaceae* family and is popularly known as Musk Willow from the Middle-East [43]. As part of a traditional medicine from ancient times, *S. aegyptiaca* is used as a confectionary, flavourful syrup and fragrance additive. The extract from bark and leaves have shown to exert beneficial effects, as laxative, cardioprotective, nervonic, sedative, hypnotic, somnolent, aphrodisiac, orexoiogenic, carnative, gastroprotector, anthelmintic and vermifuge [44]. Important findings associated with *salicaceae* family is that they contain salicylate composites such as salicylic acid which subsequently led to the finding of acetylsalicylic acid identified as aspirin, a worldwide analgesic, antipyretic, anti-inflammatory drug [45].

Later, the presence of other polyphenols such as gallic acid, caffeic acid, vanillin, p-coumaric acid, myricetin, catechin, epigallocatechin gallate, rutin and quercetin were confirmed to contribute as to the beneficial effects of *S. aegyptiaca* [46].

With the body generating OS through the Fenton and Haber Weiss leading to the initiation and development of several health complications such as diabetes, Alzheimer's disease, atherosclerosis, cardiovascular problems and various kinds of cancers [47–49] and considering the wide range of medicinal applications of *S. aegyptiaca*, the interesting and essential component in delineation of the bioactivity of its flavonoid and phenolic phytochemicals other than salicylates, is astounding.

Invariably, plant natural compounds with antioxidant activity are likely to preserve redox homeostasis disturbed during the natural cellular mechanism or the consequences of exposure to detrimental chemical agents. By rummaging for the free radicals, influencing the antioxidant non-enzymatic and enzymatic defense systems as well as drug metabolizing enzyme systems, *S. aegyptiaca* proves influential in the attenuation of OS and lipid peroxidation. These phytochemicals are expectable to modify the diverse biological activities such as inflammation, necrosis and carcinogenesis leading to cytoprotecting of cellular environment [50].

The interdependency of redox-potential, antioxidant activity and anti-inflammatory activity of gallic acid, quercetin, rutin and vanillin as well as acetylsalicylic acid has been examined [50]. Finding the relevance of the biological systems, the influence of gallic acid and acetylsalicylic acid have been studied on the drug metabolizing phase I and phase II enzymes as well as on endogenous antioxidant enzymes and peroxidative damage in the liver of C57BL/6 mice.

Oxidation–reduction potential of gallic acid, acetylsalicylic acid, rutin, quercetin and vanillin have been tested with the agents exhibiting reduction potential in a dose dependent manner of 5–50 μg/ml [50]. The reduction potential was in the order of gallic acid > quercetin > rutin > vanillin > acetylsalicylic acid.

In red and yellow onion [51] and in *S. aegyptiaca,* the antioxidant activity of gallic acid, acetylsalicylic acid, rutin, quercetin and vanillin have been examined for their scavenging of 2,2-diphenyl-1-picrylhydrazyl (DPPH) radicals and displayed inhibition of DPPH radicals, an indicator of antioxidant activity, in concentration dependent manner. In the *S. aegyptiaca* experiments, DPPH radical scavenging activity was shown by gallic acid to be greater than that of quercetin which was greater than of rutin which was greater than that of vanillin which was greater than that of acetylsalicylic acid [50].

4.1.1 Salix aegyptiaca *anti-inflammatory and antioxidant activity*

To determine the anti-inflammatory activity of the *S. aegyptiaca* phytochemicals (62.5–1000 μg/ml) inhibition of protein denaturation was used. The phytochemicals' concentration-dependent inhibitory effect displayed a relative repressive effect in the ensuing order of acetylsalicylic acid > gallic acid > rutin > quercetin > vanillin [50]. Also, cumulative therapeutic effects of phytochemicals in *Arnica montana* flower extract has been reported to alleviate collagen-induced arthritis while inhibition of both pro-inflammatory mediators and OS [52].

4.1.2 Salix aegyptiaca *and protein carbonyl estimation and oxidative stress*

Protein carbonyl measurements provide a sensitive index of OS damage occurring early in severe sepsis and major trauma patients. Elevated protein carbonyl concentrations in plasma and in bronchial aspirates indicates wide spread of oxidation though out the body beyond the lungs. The correlation between oxidative biomarkers and myeloperoxidase concentrations correlations in the lung may indicate that neutrophil oxidants could be responsible for the lung injury [53, 54] and also, protein carbonyl as a marker of OS is associated with overhydration, sarcopenia and mortality in hemodialysis patients [55]. Moreover, plasma protein carbonyls have been shown to be a predictive biomarker of oxidative stress in chronic kidney disease, dialysis, and transplantation [56]. However, Mkhwanazi *et al*. have reported that a maslinic acid triterpene derivative improved the renal function of streptozotocin-induced diabetic rats [37]. Furthermore, Mavondo et al. chronologies how malarial inflammation-driven pathophysiology and were reduced by triterpene application in various *in vivo* and *ex vivo* experiments of Asiatic acid [34].

Phytotherapeutics *S. aegyptiaca* obtained gallic acid, acetylsalicylic acid, rutin and quercetin (62.5–1000 μg/ml) showed a dose-dependent protection against protein carbonyl damage caused by the Fenton reagent with higher protection percentage as compared to the control, being maximum for gallic acid > quercetin > rutin > acetylsalicylic acid [50].

4.1.3 Salix aegyptiaca *and peroxidative damage and its inhibition*

Some water extractable phytochemicals inhibited Fe^{2+}-induced *in vitro* lipid peroxidation in a rat's brain [57] while Solanum *xanthocarpum* root extract protective efficacy was demonstrated against free radical damage and its phytochemical analysis was carried out and antioxidant effect determined [58]. Over more, microsomes were used *in vitro* to study the peroxidative damage and its inhibition by phytochemicals (62.5–1000 μg/ml) [50]. In these experiments, the Fenton reagent initiated by peroxidation and determined in terms of TBARS formation. Ultimately, all the phytochemicals showed inhibitory effect against peroxidative damage, in a dose dependent manner showing inhibition order: gallic acid > quercetin > rutin > acetylsalicylic acid.

Treatment with 50 μg/kg of acetylsalicylic acid and with 100 μg/kg of gallic acid increased cytochrome P450 reductase activity (1.26-fold, $p < 0.01$) and cytochrome b5 reductase (1.45-fold, $p < 0.01$) as equated to control [50]. Treatment using a reversed concentration combination of the two phytochemicals further enhanced the enzymatic activity (cytochrome P450 reductase vs. cytochrome b5 reductase) by a 1.58-fold and 1.66-fold when a larger dose (50 μg/kg body weight) and a lower dose (25 μg/kg body weight) of acetylsalicylic acid was used, respectively, showing their antioxidant capacity [50].

4.1.4 Glutathione S-transferase and DT-diaphorase

Precise activity of glutathione S-transferase (GST) tend to be amplified by 1.92-fold ($p < 0.01$) and 2.11-fold ($p < 0.001$) when animals were treated with 50 μg/kg of acetylsalicylic acid (group III) and 100 μg/kg weight of gallic acid as compared to controls, an observation seen with. The activity of DT-diaphorase (DTD) was also observed to be significantly elevated by 1.79-fold ($p < 0.001$) and 2.36-fold ($p < 0.001$) when animals were treated with 50 μg/kg 100 μg/kg gallic acid, respectively, as compared to control group.

For both phytochemicals, a reversed concentration combination of acetylsalicylic acid and gallic acid treatment increased the activity of both enzymes showing improved anti-inflammatory [50]. Similar observation were demonstrated with thymoquinone on antioxidant enzyme activities, lipid peroxidation and DT-diaphorase in different tissues of mice with possible mechanism of action demonstrated [59]. Also, in early cancer studies similar antioxidant and anti-peroxidation effects of dietary curcumin have been shown on glutathione *S*-Transferase and the attenuation of malondialdehyde-DNA adducts in rat liver and colon mucosa [60].

4.1.5 Superoxide dismutase (SOD), catalase, glutathione reductase and glutathione peroxidase

Superoxide dismutase (SOD), catalase (CAT) and glutathione peroxidase (GPX) plays an important and indispensable role in the entire defense strategy of antioxidants as fundamental first line defense antioxidants. This is more so with reference to super oxide anion radical ($^{*}O_2$) which is ceaselessly generated in normal body metabolism, particularly through the mitochondrial energy production pathway (MEPP) [61] and is attenuated when phytotherapeutics are administered.

Significantly increases in activities of SOD, CAT, GPX and glutathione reductase have been shown to be triggered by treating animals with gallic acid and acetylsalicylic acid. Animals treated with 100 μg/kg gallic experience enhanced SOD activity (1.47-fold). Variable concentrations of acetylsalicylic acid addition raises of these enzymatic antioxidants activity even higher [50] testimony to the efficacies of these phytochemicals in fighting lipid peroxidation.

4.2 Phytochemicals neuroprotection against oxidative stress

The foremost causes of dementia include neurodegenerative diseases and ischemic stroke and all have OS as an important player in their pathophysiology [62]. By modifying the expressions of antioxidant molecules and enzymes, the

Nrf2-ARE (nuclear factor erythroid 2-related factor 2/antioxidant responsive element antioxidant) system plays an essential role in neuroprotection as the primary cellular defense against OS. However, concurrent events of overproduction of ROS and dysregulation of the Nrf2-ARE system causes harm to indispensable cell components resulting in loss of neuron structural and functional integrity. On the other hand, TrkB (tropomyosin-related kinase B) signaling which is a classical neurotrophin signaling pathway, regulates neuronal endurance and synaptic plasticity important for fundamental functions in memory and cognition. The TrkB signaling, especially the TrkB/PI3K/Akt (TrkB/phosphatidylinositol 3 kinase/protein kinase B) pathway promotes the initiation and nuclear translocation of Nrf2, and brings in neuroprotection against OS. Essentially, the TrkB signaling pathway is also known to be downregulated in brain disorders due to lack of neurotrophin support. Therefore, activations of TrkB and the Nrf2-ARE signaling system suggests a potential approach to the design of novel phytochemical therapeutic agents for brain disorders.

The association between OS and the pathogenesis of neurodegenerative diseases, brain injury and the neuroprotective effects of phytochemicals that can co-activate the neuronal defense systems orchestrates important facets of the cellular antioxidant defense and TrkB signaling-mediated cell survival systems as possible pharmacological targets for the treatment of neurodegenerative diseases.

Factors contributive to OS in the brain include excitotoxicity, cellular antioxidant system exhaustion, lipid-rich membranes, susceptibility to lipid peroxidation, and brain high oxygen demand [63]. Excess ROS causes structural and functional modifications of cellular biomolecules, including proteins, DNA, and lipids, potentially limiting neuronal function and survival. The mechanisms rudimentary to the pathobiology of neurodegenerative diseases (NDDs) remain elusive. However, indications strongly advocates a noteworthy relationship between OS and NDDs, encompassing AD and Parkinson's disease (PD) [64]. Moreover, OS contributes to the pathogeneses of secondary damage after cerebral ischemia and other brain injuries [65, 66].

The deposition of misfolded proteins, seen in major NDDs, induce inflammatory responses, promoting ROS generation and resulting in OS [67]. Furthermore, OS causes and is caused by mitochondrial dysfunction [68]. Agreed, the central role the mitochondria play in energy metabolism and the regulation of redox homeostasis, mitochondrial malfunction contributes to the pathobiology of brain disorders. Howsoever caused, when encountered, cells compensate for the OS detrimental effect by triggering intracellular antioxidant defense system, unfortunately, contextually compromised in NDD. Therefore, activating the endogenous defense system by actuating Nrf2 using phytotherapeutics might provide a means of suppressing OS mediated cellular damage [62]. However, while OS may harm neuronal cytoarchitecture and restraining the detrimental effect of ROS alone may not suffice to prevent/reverse OS-mediated cellular damage. Approaches that support regeneration of damaged neuronal structures are necessary such that phytochemical interventions may be used for this purpose with outstanding results.

Physiologically, neuronal growth and survival are preserved via the neurotrophic signaling pathway, but modification in the regulation of specific neurotrophic factors and their receptors supervenes in the degenerating and aging brains [69]. Particularly, the brain-derived neurotrophic factor (BDNF)-dependent TrkB pathway, which is a critical signaling.

pathway for the survival and normal functioning of mature neurons, is compromised due to lack of BDNF [70, 71]. Put together, the TrkB pathway and the Nrf2 signaling system seem to suggest potential targets for encouraging neuronal survival and initiating the regeneration of injured neuronal structures and synaptic connectivity. Therefore, phytochemicals and other natural products can directly scavenge oxygen free radicals and boost the expressions of cellular antioxidant enzymes and molecules [36, 72]. This way, protection against OS-mediated cellular injury by these molecules may be possible [73, 74]. The neuritogenic potentials of the phytochemical therapeutics agents have been demonstrated [75, 76] to support the reconstruction of synaptic connectivity by renewing damaged neuronal processes [77, 78].

Different varieties of natural pharmacological modulate co-activate antioxidant defense and neurotrophin signaling-mediated cell survival systems [79–81] signifying that these compounds have therapeutic potential for the treatment of OS-mediated brain disorders. Targeting both of these signaling systems with a single compound offers benefits over combinations through possible the bypass of drug–drug interactions that could be either synergistic or antagonistic [62]. Furthermore, a single compound which can activate both the signaling and defense systems, would be more convenient to establish a therapeutic agent regarding pharmacokinetics and drug delivery.

4.2.1 Oxidative stress in case of neurodegenerative disease or brain injury

Disorders of dementia, to include NDDs, ischemic stroke or traumatic brain injury (TBI, complications are foremost public health concerns intimately linked to OS. Significantly higher concentrations of OS biomarkers and lower amounts of antioxidant biomarkers have been observed in the brain, peripheral tissues and body fluids of patients with brain disorders during preclinical and clinical studies [82, 83]. In these cases, high lipid peroxidation biomarkers are displayed as well.

4.2.1.1 Alzheimer's disease and lipid peroxidation effects

The major cause of dementia and most common progressive NDD is Alzheimer's disease [84, 85]. The main pathological hallmarks of AD, include extracellular β amyloid (βA) plaque deposits, intraneuronal aggregation of neurofibrillary tangles (NFTs), and brain atrophy [86]. Furthermore, OS has been shown to provoke βA deposition (plaque formation), tau hyperphosphorylation (NFT formation), and the ensuing degenerations of synaptic connectivity and neurons by damaging the protein degradation system [85].

Wojsiat et al., and Youssef, P., (2018) have reported raised levels of ROS-mediated vagaries in AD brains, supporting the notion that OS is caught up in the pathobiology of AD [87, 88], as shown by elevated concentrations of MDA and 4-HNE (lipid peroxidation biomarkers) being higher than normal in the brain tissues and cerebrospinal fluid samples of AD patients [84, 89]. Activities of antioxidant enzymes, superoxide dismutase (SOD), glutathione peroxidase (GPx), catalase (CAT), and peroxiredoxin (Prdx) were altered in the brain affected areas although 4-HNE levels remained unaffected [88].

Male AD patients display elevated plasma concentrations of protein carbonyls and advanced glycation end products (carboxymethyllysine and carboxyethyllysine) [90]. Furthermore, 3-nitrotyrosine (3-NT), a protein nitration product, tend to be increased in CD3C (+) T-cells from AD patients [91]. Plasma antioxidants (uric and bilirubin) are significantly decreased concurrently with reduced activities of antioxidant enzymes in AD patients [92].

Oxidative stress contributes to mitochondrial dysfunction and cellular atrophy [93] while pathological aggregations of proteins such as Aβ and tau have been reported to target mitochondria and augment ROS production [94]. OS also retards synaptic plasticity contributing to progressive memory impairment, a distinguishing clinical symptom of AD [95]. The connection between OS and AD strongly may suggest that approaches linked to antioxidant or antioxidant defense system such phytotherapeutics use could play imperative roles in the future management of AD.

4.2.1.2 Parkinson's disease (PD) and OS

The prevalence of PD is preceded only by that of AD being characterized by dopaminergic neuron degeneration in the substantia nigra [82]. A major pathological hallmark of PD is the intraneuronal aggregation of a-synuclein and the formation of Lewy bodies [93]. Crucial participation of OS in PD is intoned by convincing evidence although the exact mechanisms underlying the pathophysiology of this disease remains indescribable [96]. Singh et al., (2019) reported elevated concentration of oxidative damage markers and low concentration of glutathione (GSH) in the substantia nigra of PD patients [93]. Furthermore, high MDA plasma concentrations [97] and elevated protein carbonyl and 8-OHdG (markers of oxidative impairment to protein and DNA, respectively) in brain tissues have been reported [98]. Also, elevated concentrations of 8-OHdG and MDA, reduced activity of catalase and concentration of uric acid, and GSH have been reported in the blood of PD patients [99]. The involvement of OS in the pathobiology of PD and suggestion that targeting OS and lipid peroxidation offers a potential phytochemical therapeutic strategy for addressing this devastating brain disorder is supported.

4.2.1.3 Ischemic stroke and oxidative stress leads to lipid peroxidative damage

A sudden interruption in brain blood supply due to vascular occlusion results in a stroke which is the second leading cause of death [100] and an important source of permanent disability in adults worldwide [101]. Resultantly, a portion of the brain experiences oxygen and nutrient insufficiencies, which causes depolarization of neuronal membranes and glutamate surge into synapses, resulting in a cascade of events, including calcium overload, dissipation of mitochondrial membrane potentials, OS, and inflammation [63, 102].

Inappropriate concentrations of antiapoptotic proteins [e.g., Bcl-2 (B-cell lymphoma 2)] and proapoptotic proteins [e.g., Bax (Bcl-2-associated X protein)] contribute to mitochondrial dysfunction and OS induced apoptosis [103]. Moreover, the reestablishment of blood supply immediately after ischemia exposes brain tissue to excess oxygen, which exacerbates ROS production and in turn, induces further OS-associated injury, lipid peroxidation, protein oxidation, and intracellular DNA damage [63, 104]. After ischemic stroke, oxidative damage follows with elevated OS biomarkers (NO and MDA) concentrations reported [105]. These findings indicate targeting OS and inhibiting lipid peroxidation offers a promising therapeutic strategy to reduce secondary brain injury after ischemic stroke with possible outcomes improvement [63].

4.2.1.4 Traumatic brain injury builds oxidative stress and lipid peroxidation

Traumatic brain injury (TBI) is a major cause of death and disability world over. Non-fatal TBI may lead to neurological deficits due to direct tissue damage (primary injury) or subsequent biochemical changes (secondary injury) [106]. Biochemical factors such as excitotoxicity, inflammation, mitochondrial dysfunction, and OS drive progressive neuronal degeneration in secondary damage [107]. Importantly, further damage need be reduced by targeting the secondary changes. Indications that TBI results in OS are observed by OS biomarkers (oxidized protein moieties, lipid peroxidation products, DNA damage products) accumulating in the brain with antioxidant molecules concentrations and enzymes activities (GSH, GPx, glutathione reductase (GR), glutathione S-transferase (GST), SOD, and CAT) decline [66]. Phytochemical neuroprotective strategies, directed at salvaging injured brain tissue soon after injury and that promote regeneration during the recovery stage, are beneficial [108]. Therapeutic potentials of BDNF and its analogues have been reported in TBI and other neurological conditions [108, 109]. Therefore, phytotherapeutics targeting cellular antioxidant defense and the BDNF/TrkB signaling pathway might improve cognitive deficits secondary to TBI.

5. Phytochemicals that activate neuronal antioxidant defense and survival mediating against lipid peroxidation

Many plant-derived bioactive molecules deactivate ROS and reportedly, potentiating the cellular antioxidant system. The principle of action of inducing an antioxidant effect by promoting adaptive cellular stress response using phytochemicals is substantially supported [110]. Furthermore, phytochemicals have been shown to protect neurons from OS by activating TrkB signaling pathways and the Nrf2-ARE system promoting cellular survival [77, 78].

5.1 Phenolic compounds and OS in brain neurological damages

5.1.1 Sulfuretin quenches oxidative stress effects

Numerous phenolics exhibit neuroprotective effects against OS in models of AD and other neurodegenerative disorders. Sulfuretin (**Figure 2**), a flavonoid glycoside isolated from the stem bark of *Albizia julibrissin* and heartwood of *Rhus verniciflua*, protected SH-SY5Y cells and primary hippocampal neurons from Aβ-induced neurotoxicity [111]. The PI3K/Akt and Nrf2/HO-1 signaling pathways may contribute to sulfuretin-mediated neuroprotection through inhibiting cell death by suppressing ROS production, enhancing PI3K/Akt pathway and the nuclear translocation of Nrf2 (**Figure 3**) [62]. Also, the phytochemical sulfuretin was shown to suppress adipocyte differentiation of preadipocytes and prevented obesity and increased insulin sensitivity by suppressing expression of inflammatory markers, inducing expression of adiponectin, and increasing concentrations of phosphorylated ERK and AKT [112]. Using a microarray analysis and identification of activating

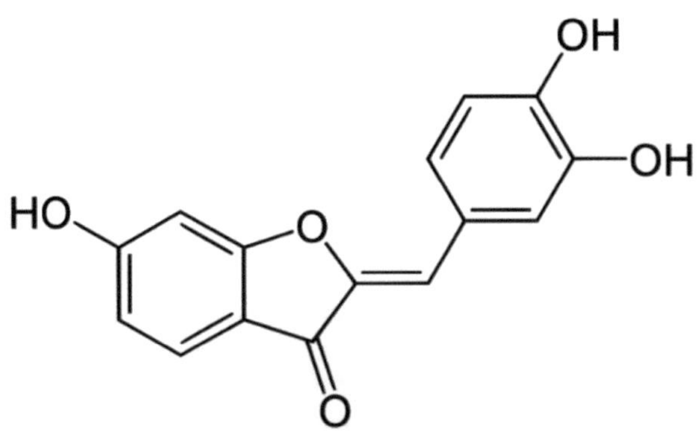

Figure 2.
Sulfuretin.

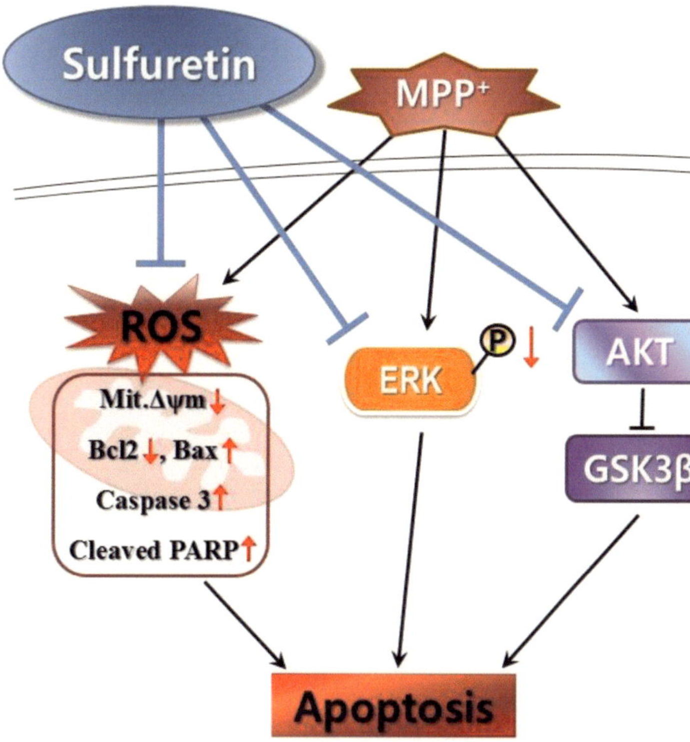

Figure 3.
Other inhibitory effects of sulfuretin. ERK- extracellular signal-regulated kinase; AKT- protein kinase B (PKB); GSK3β- glycogen synthase kinase 3 beta; PARP = poly (adenosine diphosphate-ribose) polymerase; MPP^+- 1-methyl-4-phenylpyridinium. ⊣ inhibition; ↑ increases; ↓ decreases. MPP^+ arises as the toxic metabolite of the compound MPTP. MPTP is converted in the brain into MPP^+ causing parkinsonism killing dopamine-producing neurons in the substantia nigra through oxidative stress generation which is inhibited by sulfretin.

transcription factor 3 (Atf3) as a sulfuretin-responsive gene, the molecular mechanism of sulfuretin in adipocytes was illuminated. Administration of sulfuretin raised the Atf3 mRNA and protein concentrations in white adipose tissue and adipocytes. Reliably, Atf3 deficiency promoted lipid accretion, the adipocyte expression and lipid peroxidation markers. Sulfuretin's but not resveratrol's anti-adipogenic effects were diminished in cell with Atf3 deficiency, indicating that Atf3 is an essential factor in the function of sulfuretin [112] and that phytochemicals are essential for as antioxidant agents.

5.1.2 Resveratrol has antioxidant effects in brain disease

By activating the PI3K/Akt/Nrf2 pathway, the grape polyphenol and anthocyanin, Resveratrol (**Figure 4**), and a derivative of Korean black beans, protected PC12 cells [81] and HT22 cells [113], respectively, against Aβ-induced toxicity. In Aβ-induced toxicity, resveratrol inhibited cell death and suppressed OS markers such as MDA and ROS by elevation of the phosphorylation of PI3K and Akt, the nuclear translocation of Nrf2, and the protein concentrations of SOD, HO-1, and GSH [81]. Anthocyanins diminished cell death by modifying the expressions of proapoptotic markers (cleaved caspase-3) and stress markers (MDA, H_2O_2, 8-OHdG) while enhancing the glycogen synthase kinase-3 beta (GSK3b), phosphorylation of PI3K and Akt, the nuclear translocation of Nrf2, the expression of HO-1, and GSH levels.

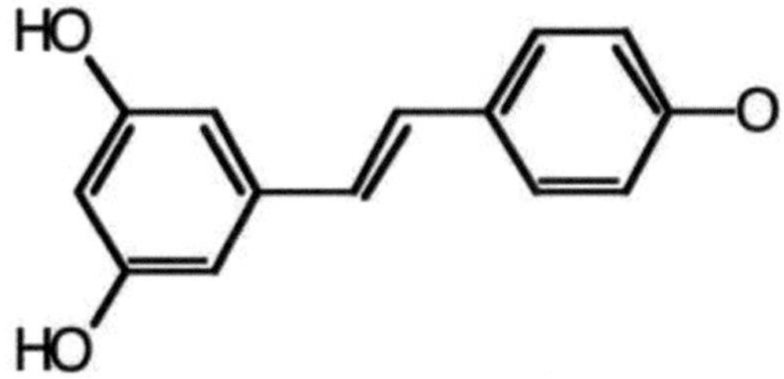

Figure 4.
Resveratrol whose chemical name is 3,5,4′-trihydroxy-trans-stilbene is a stilbenoid, a type of natural polyphenol, and a phytoalexin synthesized by numerous plants in response to injury or when the plant is under attack by pathogens, such as (viruses, bacteria or fungi).

5.1.3 Tea polyphenols, oxidative stress, neuroprotection

Tea polyphenols (TPs) attenuated OS in H_2O_2-stimulated SH-SY5Y cells by activating the Keap1-Nrf2 signaling pathway and the TrkB/CREB/BDNF pathway [114]. Also in these experiments, TPs attenuated H_2O_2-induced cell death, mitochondrial dysfunction and reduced elevated ROS and H_2O_2 concentrations. Moreover, TPs heightened the nuclear translocation of Nrf2 and the TrkB/CREB/BDNF signaling mechanism by activating the PI3K/Akt pathway, and thus, transcriptionally regulating the downstream expressions of HO-1, NQO1, SOD, GPx, and CAT in SH-SY5Y cells [114].

5.1.4 Isoflavone of fermented soy and neuroprotection

8-Hydroxydaidzein (8-OHD), an isoflavone of fermented soy, protected against neuroinflammation in LPS-stimulated BV2 microglial cells by stimulating Nrf2-antioxidant and Akt/NF-kB inflammatory signaling pathways. In BV2 microglial cells, 8-OHD inhibited the LPS-activated productions of NO, TNF-a, and IL-6 by suppressing gene expression [115]. Moreover, 8-OHD quenches ROS and promotes the nuclear translocation of Nrf2, and thus, upregulates the expressions of Phase II enzymes, such as HO-1, NQO1, and GCL [115]. 8-OHD also suppresses the LPS-stimulated phosphorylation of Akt and NF-kB-p65 attenuating LPS-induced prostaglandin E2 (PGE2) production without affecting COX-2 expression [116].

5.1.5 Rutin protection against neurotoxicity

A flavonoid found in buckwheat, rutin protected male albino SD rats from acrylamide or g-radiation-induced neurotoxicity by activating the PI3K/Akt/GSK-3b/NRF-2 signaling pathway (**Figure 5**) [118]. Rutin (**Figure 5**) increases the phosphorylation of PI3K, Akt, and GSK-3b and the nuclear translocation of Nrf2, suppressed MDA levels, GST activity, and the expressions of IL-1b and IL-6, and increased IGF1 and NGF levels [118]. The phytochemical rutin's mechanisms of action includes reduction of proinflammatory cytokines, increasing antioxidant enzyme activities, stimulation of the mitogen-activated protein kinase cascade, downregulation of mRNA expression of PD-linked and proapoptotic genes, upregulation of the ion transport and antiapoptotic genes, and restoration of the functions of mitochondrial complex enzymes [119]. Taken together, these suggest that they phytotherapeutic rutin may be a hopeful neuroprotective compound for the treatment of NDDs.

Rutin as neuroprotective agent has been seen to move from bench to the bedside as a phytotherapeutic [120] as well as the inhibition of neuroinflammation and

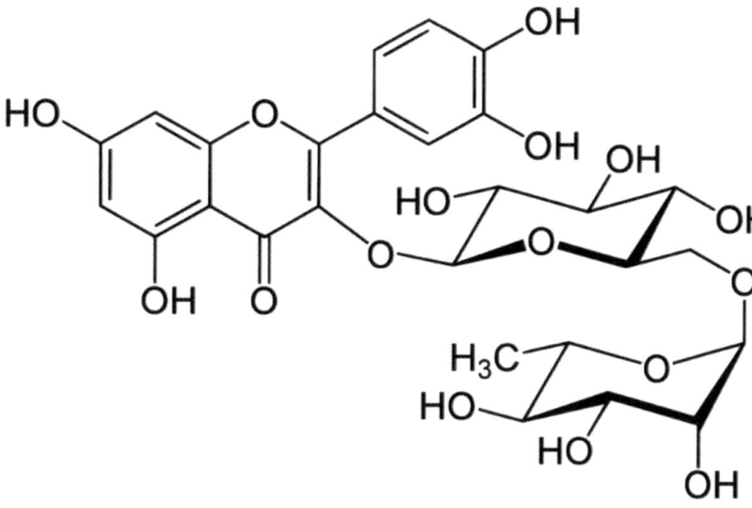

Figure 5.
Rutin, also called rutoside, quercetin-3-O-rutinoside and sophorin, is the glycoside combining the flavonol quercetin and the disaccharide rutinose (α-L-rhamnopyranosyl-(1 → 6)-β-D-glucopyranose) [117].

providing neuroprotection in subarachnoid hemorrhage through suppressing the RAGE-NF-κB inflammatory signaling pathway [121].

Rutin has anti-inflammatory, antioxidant, anti-viral, anti-tumor and immune regulatory effects. Neuroprotective mechanisms of rutin for spinal cord injury has been reported to occur through anti-oxidation and anti-inflammation and inhibition of p38 mitogen activated protein kinase pathway [122].

5.2 Non-phenolic compounds, neuroprotection and lipid peroxidation

Numerous non-phenolics have been reported to exhibit neuroprotective effects against OS in models of AD and other neurodegenerative disorders [62].

5.2.1 Brassicaphenanthrene A, acerogenin A and neuroprotection

Isolated from *Brassica rapa*, Brassicaphenanthrene A protects HT-22 neuronal cells from glutamate-induced excitotoxicity and upregulates Nrf2-mediated HO-1 expression via PI3K/Akt and JNK regulatory pathways [123]. Acerogenin A, an isolate from the stem bark of Acer *nikoense* (Japanese traditional medicine) protected HT22 cells from glutamate-induced oxidative injury [124] through the stimulating the PI3K/Akt/Nrf2/HO-1 pathway. Acerogenin A diminished cell death by suppressing the production of ROS and increasing the nuclear translocation of Nrf2, the expression of HO-1, and the phosphorylation of Akt [124].

Polysaccharide extracts (PPE) of *Perilla frutescens* triggered PI3K/Akt and Nrf2-mediated HO-1/NQO1 pathways and protected against H_2O_2-induced OS in HT22 cells [125]. The PPE attenuates cell damage by suppressing the expressions of Bax, cytochrome C, and caspases-3,-8, and – 9, and enhancing the expressions Bcl-2 and Poly [ADP-ribose] polymerase (PARP) while increasing MAPKs (p38, ERK, JNK), PI3K, Akt and p65 phosphorylation, decreasing NF-kB concentration and enhancing the nuclear translocation of Nrf2 and the expressions of HO-1 and SOD [125].

3,3'-Diindolylmethane, a metabolite of indole-3-carbinol found in Brassicaceae family, attenuates OS in glutamate-induced HT-22 cells by activation of the TrkB/Akt pathway [126]. 3,3'-Diindolylmethane metabolite moderates the expressions of Bax, cytochrome c, cleaves caspase-3, and AIF (apoptosis-inducing factor), and improves Bcl-2 expression, the phosphorylation of TrkB, Akt, and CREB, and the expressions of HO-1, GCLC, NQO-1, and GPx. Scopolamine-treated mice improved cognitive deficits after 3,3'-Diindolylmethane administration [126].

Diallyl trisulfide, an organosulfur compound in garlic oil, stimulated the PI3K/Akt-mediated Nrf2/HO-1 signaling pathway and protected against OGD-induced neuronal injury [127]. Also, diallyl trisulfide inhibited the expressions of proapoptotic markers (cleaved caspase-3), OS markers (ROS and MDA), and increased the nuclear translocation of Nrf2, the expression of antioxidant enzymes (e.g., HO-1), and the phosphorylation of Akt [127] thus reducing lipid peroxidation potentials in brain injury.

Oxymatrine, isolated from the Chinese herb *Sophora flavescens* protected P7 SD rats from hypoxic–ischemic brain injury [128], through activating the Akt/GSK3b/HO-1/Nrf-2 signaling pathway. Moreover, oxymatrine increased the nuclear translocation of Nrf2, the phosphorylation of Akt and GSK3b, and HO-1 expression and attenuated the degree of neurological deficits [128].

6'-O-Galloylpaeoniflorin, a galloylated derivative of paeoniflorin isolated from peony root, protected an OGD-induced ischemic PC12 cell model and a CIRI male Wistar rat model against ischemic stroke through stimulating PI3K/Akt/Nrf2 pathway [129]. Also, 6'-O-galloylpaeoniflorin attenuated OS and neuroinflammation while improving neurological deficits, inhibiting apoptosis by suppressing the expressions of pro-apoptotic markers (cleaving caspase-3), inhibiting inflammatory cytokine (TNF-a, IL-1b), and MDA concentration. Nuclear translocation of Nrf2 and SOD expression increased through elevation of Akt phosphorylation [129].

Ginkgolides A, B and C are diterpenes isolated from *Ginkgo biloba* L. and defends PC12 cells from OGD/R-induced ischemic injury in adult male SD rats subjected to MCAO-induced acute cerebral ischemic injury [130]. They stimulate Akt/Nrf2 and Akt/CREB signaling pathways. These ginkgolides inhibit cell death by overwhelming the expressions of Bax and cleaved caspase-3, enhancing the phosphorylation of Akt and pCREB, and increasing the nuclear translocation of Nrf2 and HO-1 expression [130]. Also, ginkgolides protects against ischemic stroke in an OGD-induced SH-SY5Y cell ischemic model and MCAO-induced model of cerebral ischemic injury in male SD rats [131]. Ginkgolides suppress ROS production and increase Akt phosphorylation, the nuclear translocation and phosphorylation of Nrf2, and the expressions of HO-1, Nqo1, and SOD [131]. In the process lipid peroxidation associated damages are averted.

PC12 cells are protected against OGD/R-induced neuronal injury by **Protodioscin** which actuates the PI3K/Akt/Nrf2 pathway. This occurs through elevating the expressions of HIF-1a, SOD, GPx, HSP70, and HO-1, the phosphorylation of PI3K and Akt, the nuclear translocation of Nrf2, and upregulating miR-124. As a result, OS is attenuated [132] and lipid peroxidation reduced.

Matrine, a quinolizidine alkaloid, derived from the herb Radix Sophorae *flavescentis*, protected rats from subarachnoid hemorrhage [133] through PI3K/Akt-mediated NF-kB inhibition and Keap1/Nrf2-dependent HO-1 induction. The effects of matrine included inflammatory cytokines (TNF-a, IL-1b) and pro-apoptotic markers (Bax and cleaved caspase-3) expression suppression with enhanced pro-survival marker Bcl-2 ([133]. Also, matrine increased nuclear translocation of Nrf2 and HO-1 expression and lowered NF-kB P65 expression by increasing the phosphorylation of Keap1, Akt, and IkB-a [72] invariably attenuating lipid peroxidation.

Panax notoginseng saponins protected against blood–brain barrier (BBB) injury [134] by activating the PI3K/Akt/Nrf2 antioxidant signaling pathway. In LPS-stimulated cerebral microvascular endothelial cells BBB injury model, saponins attenuated the creations of ROS and inflammatory cytokines (ILβ1b, TNFα), decreased NFβkB levels, and increased the nuclear translocation of Nrf2 HO-1 expression, and the phosphorylation of Akt [134].

6. Lipid peroxidation-induced inflammation and oxidation-mediated degenerative diseases

Degenerative disease is positioned as one of the most fatal group of diseases contributing to the mortality, poor quality of life, increasing economic problems of the sufferers with OS and inflammation being leading drivers of lipid peroxidation related pathology [135]. The most common degenerative diseases include rheumatoid arthritis (RA) [136], diabetes mellitus (DM) [137], and cardiovascular disease (CVD) [138]. Although a number of synthetic medications are used to treat these diseases, none of the current regimens are completely safe. Phytochemicals: polyphenols, carotenoids, anthocyanins, alkaloids, glycosides, saponins, and terpenes are potential sources of alternative medications to attenuate the oxidative stress and inflammation associated with degenerative diseases. Some of these active compounds have shown good promise for development into novel mediators for treating RA, DM, and CVD by targeting OS and inflammation.

6.1 Phytotherapeutics and degenerative diseases

Several synthetic regimens are used to attenuate oxidative stress and inflammation-mediated degenerative diseases, with most of them exerting considerable side effects when utilized in the treatment of CVDs [139], DM [140] and RA [141].

6.1.1 Cardiovascular diseases, oxidative stress and lipid peroxidation

CVDs are a group of diseases associated with complications of the heart and blood vessels invariably leads to coronary heart disease a major component of CVDs [135]. Lipid peroxidation associated major risk factors of CVDs include hypertension (HTN), hypercholesterolemia, diabetes, obesity, inflammation, smoking, consumption of alcohol, lack of exercise, and a familial history of heart diseases [142].

6.1.2 Pathogenesis of CVDs

Atherosclerosis occurs due to the accumulation of atherosclerotic plaques within the walls of the arteries is the major precursor of CVDs. Plaque formation originates from endothelial damage, followed by adherence of circulating monocytes and subsequent exposure to homocysteine, inflammation, increased platelet aggregation, and higher levels of oxidized low-density lipoprotein (LDL-ox) and ROS [143]. Moreover, increased serum triacylglycerols.

(TAG), cholesterol (C), increased plasma fibrinogen and coagulation factors, hyperglycemia, HTN and lipid peroxidation are crucial in the pathogenesis of CVDs [144].

6.1.3 Phytochemicals and CVDs

Polyphenols and other antioxidants from fruits, vegetables, and spices has the potential to lower CVD risks by attenuating oxidative stress and inflammatory mediators [135]. Fruit consumption in Japan protected against the risk of CVDs [145], a higher consumption of fruits and vegetables correlated with a lower risk of all-cause mortality, predominantly cardiovascular mortality [146] while high-frequency consumption of fruits and vegetables lowered plasma C-reactive protein (CRP) and homocysteine concentrations, accordingly reducing inflammation a considered high risk factor of CVDs [147]. High fruit consumption level decreased HTN and blood glucose concentration which significantly decrease the risks of CVDs [148].

Polyphenolic extract from the apple has a significant effect on decreasing the serum total-C and LDL-C levels in healthy individuals with relatively high body mass index (BMI), which consequently limits CVD risk [149]. Consumption of banana decreases the oxidative modification of LDL, plasma lipids, and lipoproteins and thus ultimately aids in protection from atherogenesis due to its antioxidant properties [150]. Furthermore, blueberries, strawberries, and cranberries reduce cardiovascular risk factors of lipid peroxidation, inflammation, and regulate HTN due to the presence of high concentrations of anthocyanins and ellagitannins in their skin and flesh [151, 152]. Moreover, being good sources of polyphenols, berries have a high content of micronutrients such as folate, α-carotene, β-carotene, potassium, vitamin C, and vitamin E, which exhibit noteworthy antioxidant activities [153].

Citrus fruits such as mandarins, lemons, oranges, and grapefruits contain high quantities of flavanones (naringin and hesperidin stimulate nitric oxide in endothelium) that improve significant vascular functions and the lipid profile in coronary artery diseases patients [154, 155]. Pomegranate fruit juice and peel extracts have antihypertensive, anti-atherosclerotic, antioxidant, and anti-inflammatory effects because of polyphenolic compounds including anthocyanins, catechins, and tannins, contributing to the attenuation of CVD risk factors [156].

Polyphenol-rich peach and plum juice prevent against risk factors effects for cardiometabolic disorders by decreasing the expression of plasma proatherogenic and proinflammatory molecules, intercellular cell adhesion molecule-1 (ICAM-1), monocyte chemotactic protein-1, and nuclear factor kappa B (NF-κB) and by decreasing foam cell adherence to aortic arches. Furthermore, peach and plum juice reduce plasma angiotensin II activity and the expression of its receptor Agtr1 in cardiac tissues. Peach and plum polyphenols act as peroxisome proliferator activated receptor-γ (PPARγ) agonists [157]. An *in vivo* and *ex vivo* experiment watermelon improves lipid profiles and antioxidant capacity and decreases inflammation and alters gene expression for lipid metabolism and consequently reduce CVDs risk factors [158].

Sulfur-containing organic compounds (organosulphur) from garlic (*Allium sativum*), onion (*Allium cepa*), and cruciferous vegetables such as broccoli, cauliflower, cabbage, and Brussels sprouts exhibited cardioprotective effects facilitated by their antioxidant and anti-inflammatory properties [159]. Garlic's sodium 2-propenyl thiosulfate is suggested to block platelet aggregation through inhibition of ADP and platelet-activating factor (PAF) [160]. The onion key flavonoid, quercetin (3,3′,4′,5,7-pentahydroxyflavone), has anti-atherosclerotic properties and accumulate in the aorta tissue where its metabolites exert antioxidant and anti-inflammatory activities [161]. The bright red carotene and carotenoid pigment in tomato, lycopene, significantly reduces myocardial infarction (MI) in isoproterenol injected rats [162]. Supplementation of tomato and corn oil improves diastolic function, changes cardiac miRNA expression, and attenuates lipid hydroxy peroxidation and oxidative stress [163].

Ginger *(Zingiber officinale)* benefits in the treatment of CVDs exhibiting anti-inflammatory as well as antithrombotic properties by inhibiting the production of NO, inflammatory cytokines, cyclooxygenase (COX), and lipoxygenase (LOX) with no or very few side effects as compared to nonsteroidal anti-inflammatory drugs (NSAIDs) [164, 165]. Ginger is known for its being an antioxidant, antiplatelet aggregation, positive inotropic, hypotensive, hypoglycemic and hypolipidemic in *in vitro, in vivo* studies and in human clinical trials [166].

Black pepper, and its active ingredient (piperine) influences significant decrease in the concentrations of free fatty acids, phospholipids, and triacylglycerols and an increase in the concentration of high density lipoprotein cholesterol (HDLC), thus reducing the risk of atherosclerosis [167, 168].

7. Diabetes mellitus and lipid peroxidation

7.1 Pathogenesis from inflammation and lipid peroxidation

The pathogenesis of DM is closed associated with involvement of low-grade chronic inflammation and the activation of the innate immune system [169]. Excessive concentrations of glucose and free fatty acids initiate cellular OS in the pancreatic islets and insulin-sensitive tissues including adipose tissue, leading to the activation of c-Jun N-terminal kinase (JNK) and NF-κB [170]. Increases in the adipocyte proinflammatory cytokines production including tumor necrosis factor alpha (TNF-α), interleukin (IL) 6, IL-1β, leptin, resistin, and chemokines such as MCP-1, CC-chemokine ligand 2 (CCL2), CCL3, and CXC-chemokine ligand 8 occurs when inflammatory signaling pathways start regulating protein phosphorylation and cellular transcriptional events. Accordingly, recruitment to immune cells such as monocytes to the adipose tissues contributes to tissue inflammation. Differentiation of monocytes into macrophages creates several inflammatory cytokines, further encouraging local inflammation. Moreover, the release of cytokines and chemokines from the adipose tissues into the circulation promotes inflammation in other tissues including the pancreatic β-islets [170] worsening diabetes mellitus status.

Inflammation-induced insulin resistance build up is further escalated by JNK and IKKβ/NF-κB which play important roles in inflammation. The stress kinase, JNK, normally phosphorylates the c-Jun component of the AP-1 transcription factor promoting insulin resistance. Phosphorylation of the serine residues in the insulin receptor substrate 1 (IRS-1) is involved [171]. Subsequently, counter-regulatory serine/threonine phosphorylation [172] inhibits insulin receptor signaling that normally occurs through a tyrosine kinase cascade [173]. The IκB protein inhibitors of NF-κB are the highly selected physiological substrates for IKKβ. The NF-κB is inhibited by IκBα which when phosphorylated by IKKβ undergoes proteasomal degradation releasing the former for translocation into the nucleus, where it promotes the expression of numerous target genes whose products induce insulin resistance. IKKβ causes insulin resistance through the transcriptional activation of NF-κB. Therefore, decreasing gene expression and improving insulin resistance may be achieved by administration of anti-inflammatory phytochemicals. Increasing adiposity is reported to upsurge inflammatory gene expression in the liver [174], which further increases the production of cytokines and chemokines aspects ameliorated by triterpenes phytotherapeutics [34]. Immune cells including monocytes and macrophages are recruited and/or activated, which leads to local insulin resistance processes that may be averted by anti-inflammatory triterpenes like Asiatic acid, maslinic acid or oleanolic acid.

Oxidative stress modifies the enzyme systems, impairs glutathione metabolism, causing lipid peroxidation and reducing vitamin C concentration and thus contributing to DM [175]. Actually, a mutual relationship exists between hyperglycemia and oxidative stress in DM with hyperglycemia fueling glucose autooxidation, NADPH oxidase activity, oxidative phosphorylation, protein glycation, and the polyol pathway, which leads to ROS generation and OS [176]. Healthy cells are damaged functionally and structurally by ROS, losing cellular integrity leading to many pathophysiological conditions.

7.2 Phytochemical intervention in DM

Was it not for their side effects, numerous synthetic drugs groups and insulin groups possess antioxidative and anti-inflammatory potential that can be used in the treatment of DM. Resultantly, the pursuit for alternative and safer treatment regimens for DM management remains open for investigation.

7.2.1 Phytochemicals interventions in DM

In streptozotocin- (STZ) induced diabetic Sprague Dawley rats, the oral administration of **naringin** (4′,5,7-trihydroxyflavonone-7-rhamnoglucoside) at 50 mg/kg/day reduces OS and increases fasting plasma insulin activity. Naringin, the foremost flavonoid in grapefruit juice, ameliorates OS and improves ATP synthesis in pancreatic β-cell mitochondria and upgrades the subsequent insulin secretion by β-cells [177]. Significant amelioration of β-cell dysfunction, insulin resistance and hyperglycemia, reduction of TNF-α, IL-6, CRP, increase in antioxidant enzyme activities, reduced NF-κB expression, and upregulated adiponectin and PPARγ expression have been observed through the application of naringin on diabetic Wistar albino male rats for 28 days. The positive alterations were obtained with naringin treatment at 25, 50, and 100 mg/kg/day. Furthermore, naringin effectively rescues kidney cells, β-cells, and liver cells from continued pathological modifications and oxidative damage [178].

Resveratrol (**Figure 3**), a phytochemical, exerts potent antioxidative, antidiabetic, and anti-inflammatory activities. In the liver and spleen of STZ-induced male Long-Evans rats (type 1 DM), resveratrol administration (0.1 or 1.0 mg/kg/day) for 7 days, significantly decreased OS (including manganese-superoxide dismutase expression, superoxide anion content, protein carbonyl concentration). Also, reduction in hepatic inflammation factors (NF-κB and IL-1β) and decreasing the TNF-α and IL-6 concentrations in spleen were observed [178].

Apples contain a principal phenolic compound called **Phlorizin** (PZ). Preexposure of PZ- docosahexaenoic acid ester (DHA) onto a lipopolysaccharide (LPS) stimulated macrophages inflammation model effectively reduced TNF-α, IL-6, and COX-2 protein concentrations compared with DHA alone. However, both PZ-DHA ester and DHA have the potential to inhibit NF-κB activation a proinflammatory marker. Therefore, PZ-DHA ester has the potential to quench T2DM-associated inflammation [179] and ameliorate the disease.

Diabetes mellitus is associated with the glutathione concentration reduction demonstrating the critical role of OS in its pathogenesis. Pretreatment of Ins-1E pancreatic β-cells with the **flavonoid epicatechin** (present in green tea, grapes, and cocoa) prohibited tert-butyl hydroperoxide induced cell damage, ROS and p-JNK over expression. Over more, insulin secretion which indicates the protective potentiality of epicatechin against oxidative stress on β-cells is restored [180].

Pomegranate (*Punica granatum*) fruit contains flavonoids such as anthocyanins, flavonols, ellagitannins, gallotannins, and proanthocyanidins providing beneficial effect in T2DM by reducing lipid peroxidation and OS. Also effected is the increasing of enzymatic antioxidant activity, decreasing ROS, and preventing activation of PPARγ and NF-κB by pomegranate [181].

Anthocyanins, found in tart cherry, alter tissue PPARγ activity affecting metabolism and inflammation. The intake of tart cherry reduces retroperitoneal IL-6 and TNF-α mRNA expression, NF-κB activity, and plasma IL-6 and TNF-α concentrations while increasing retroperitoneal PPARα and PPARγ mRNA expression in Zucker fatty rat model of obesity and metabolic syndrome. The risk of T2DM development tend to decrease when systemic and local inflammation, metabolic syndrome and lipid peroxidation are reduced [182].

Adipose tissue LPS-induced macrophages infiltration increased adiposity and lipid peroxidation may lead to T2DM. In an *in vitro* inflammation model where the pathologic relationships between adipocytes and macrophages were mimicked, anthocyanin-rich fractions from blackberry-blueberry beverages inhibited NO and TNF-α the secretion and the phosphorylation of NF-κB p65 [183] and lipid peroxidation tendency.

T2DM is associated with chronic, low-grade, systemic inflammation accompanied by an increased production of adipokines or cytokines by obese adipose tissue. Grape fruit (0.5 g/kg/ six weeks) treatment of diabetic db/db mice produced antihyperglycemic effects that were accompanied by reduced mRNA expression of proinflammatory genes such as COX-2, monocyte chemotactic protein-1, TNF-α, NF-κB and reduced lipid peroxidation in the liver and epididymal adipose tissue [184].

The immunomodulatory effects of a mycelial submerged culture and broth of *Grifola frondosa* **mushrooms** on splenocytes and peripheral blood cells. Two weeks of intragastric administration of fermented mycelia, broth, or their combination (1 g/kg/day) into DM Wistar rats significantly decreased the 2-hour postprandial blood glucose level, the production of T-leukocyte-derived interferon gamma (IFN-γ), monocyte-derived IL-4 and IL-6, and T-splenocyte derived IL-4 which treatment significantly enhanced macrophage-derived TNF-α production [185] and possible decreased lipid peroxidation.

The administration of **fermented carrot juice** (by *Lactobacillus plantarum* NCU116) for five weeks in STZ-induced diabetic rats positively regulated the blood glucose concentration, hormone, and lipid metabolism, reestablished the antioxidant capacity, restored the morphology of pancreas and kidney, and upregulated the LDL receptor, cholesterol 7α-hydroxylase (CYP7A1), GLUT4, and PPARα and PPARγ mRNA expression [186].

In diabetic male C57BL/6 J mice, (0.5 mg/kg) treatment for 4 months with **Sulforaphane (SFN)**, an isothiocyanate found in broccoli, significantly inhibited cardiac lipid accumulation and enhanced cardiac inflammation, OS, and fibrosis. By downregulating diabetes induced PAI-1, TNF-α, CTGF, TGF-β, 3-NT, and 4-HNE expression, SFN ameliorated lipid peroxidation potential. Also, SFN, upregulated nuclear factor (erythroid-derived 2-) like factor 2 (Nrf2) and its downside genes, NQO1 and HO-1 in rescuing DM sequalae. Of note, SFN diminished 4-HNE-LKB1 adducts and reversed the diabetes-induced inhibition of LKB1/AMPK and its downstream targets, including sirtuin 1, PGC-1α, phosphorylated acetyl-CoA carboxylase, and carnitine palmitoyl transferase-1. Ultimately, SFN treatment of T2DM attenuate the cardiac OS-induced inhibition of the LKB1/AMPK signaling pathway, thereby preventing T2DM-induced lipotoxicity and cardiomyopathy [187].

Onion-derived quercetin derivatives are important flavonoids for improving diabetic conditions in both *in vivo* and *in vitro* models. Eight days of treatment with onion peel extract (1%) improved significantly glucose tolerance, liver and skeletal muscle glycogen content, and insulin receptor and GLUT4 expression in muscle tissues of STZ-induced DM in male Sprague Dawley (SD) rats. The OS-inducing dysregulations of SOD activity, increased free fatty acids plasma concentrations, the formation of MDA, and IL-6 over expression in hepatic tissue, were significantly suppressed in this model of onion-derived quercetin treatment [188].

Traditional medicinal mushroom known as *Cordyceps militaris* are a source of **Cordycepin (3′-deoxyadenosine)**. Inhibition of NO, suppression of NF-κB activation, and protein expression suppression of proinflammatory mediators that further inhibit the production of proinflammatory cytokines such as IL-1β, IL-6, and TNF-α has been demonstrated in LPS-stimulated 263.7 cells treated with cordycepin. Over more, an elevated concentration of cordycepin reduced the T2DM-regulating genes such as 11β-HSD1 and PPARγ as well as the expression of costimulatory molecules such as ICAM-1 and B7–1/−2 [189].

Curcumin (a polyphenolic compound) in Turmeric (*Curcuma longa*) is the active ingredient possesses broad-spectrum biological activities such as anti-inflammatory, antioxidant, and antitumor. Reduction of OS and inflammatory responses and inhibition of prostaglandin E2 (PGE2) and NOS have been observed in the injured lungs of DM rats after administration of curcumin. As a mode of action, curcumin inhibited the stimulation of NF-κB, a key player in inflammatory responses [190]. Eight weeks treatment of db/db mice with curcumin improved AMPK and PPARγ expression and reduced NF-κB protein levels [191].

Hyperglycemia-mediated OS of DM may induce neuronal injury. **Curcuminoids,** polyphenols of turmeric, displayed protective effects against OS in the brain of STZ-induced diabetic rats by restoring the normal concentrations of lipid peroxidation and nitrite content and endogenous antioxidant marker enzymes [192]. *De novo* synthesis of glutathione and the suppression of insulin receptor expression were achieved with the administration of curcumin which attenuated insulin-induced OS in hepatic stellate cells by stimulating the expression of glutamate-cysteine ligase [193]. Pretreatment with a novel curcumin analogue (B06) at 5 μM significantly reduced the high-glucose-induced overexpression of inflammatory cytokines in macrophages through the inhibition of c-Jun N-terminal kinase/NF-κB activation [194].

Administration of **ginger powder** (0.5%, 1%, and 5%) in STZ-induced inbred male Wistar/NIN rats for one month protected against DM effects by modulating antioxidant enzymes, glutathione and downregulating lipid and protein oxidation [195]. Combined **garlic bulb, ginger rhizome, turmeric rhizome** (200 mg/kg body weight) treatment for 28 uninterrupted days significantly alleviated hyperglycemia and dyslipidemia, increased insulin production, enhanced GSH, and decreased lipid peroxidation in nicotinamide and STZ-induced diabetic rats [196].

Diabetic encephalopathy is one of the more severe complications of DM characterized by severely reduced body weight. **Saffron** at 40 and 80 mg/kg significantly increased the body weight and serum TNF-α concentrations and decreased the blood glucose, glycosylated proteins, and advanced glycation end product (AGE) serum concentrations in DM encephalopathy rats. Additionally, saffron significantly increased the glutathione content, superoxide dismutase, and catalase but remarkably decreased the cognitive deficit and serum TNF-α, and it induced NOS activity in hippocampus tissue [197].

Administration of **Crocin,** an active constituent of saffron, significantly decreased MDA ($p < 0.01$) and xanthine oxidase ($p < 0.05$) activities while elevating glutathione ($p < 0.05$) concentration, thus ameliorating renal injury in STZ-induced rats [198]. **Safranal** is one of the components of the saffron plant which, in high-fat diet (HFD) and STZ-induced T2DM rats, treatment for a period of 4 weeks diminished OS caused by T2DM and reduced the inflammation by decreasing plasma and pancreas tissue the TNF-α and IL-1β s concentrations [199].

The protective effect of **Onion** protected against OS *in vivo* where STZ-induced male diabetic Wistar rats administered 1 mL/day of *Allium cepa* solution (0.4 g

Allium cepa/rat) improved the fasting serum HDL concentration, alleviated hyperglycemia by diminishing SOD activities [200]. Another *in vivo* study also investigated the protective effects of onion against oxidative stress; 12 weeks of onion intake suppressed the diabetes-induced oxidative stress more effectively in STZ-induced diabetic rats [140]. Onion powder (7% w/w) administration suppressed the glutathione peroxidase, glutathione reductase, and glutathione S-transferase activities [201].

Mustard leaf (*Brassica juncea*) strongly inhibits the AGE formation and free radical-mediated protein damage oral ingestion by STZ-induced diabetic rats an EtOAc fraction (50 and 200 mg/kg body weight/day/10 days) reduced the serum glucose and glycosylated protein, superoxide and nitrite/nitrate concentrations. This suggests that the EtOAc fraction of mustard leaf has the capacity to attenuate damage caused by the oxidative stress involved in diabetes and its complications [202]. Brown mustard is a high source of Asiatic acid, a triterpene with an excellent antioxidant capacity, potent anti-inflammatory [203], antihyperglycemic [204], antihyperlipidemic [35], reduction of OS while rescuing malarial infection in SD rats.

8. Rheumatoid arthritis (RA) and phytochemical interventions

RA is an inflammatory, systemic autoimmune syndrome with primary degenerating articular structures involving, the cartilage (movable synovial joints- knees, shoulders, hands) and the bones (osteoarthritis and osteoporosis) as a result of pannus development over the joint surfaces (abnormal layer of fibrovascular or granulation tissues) [205]. great socioeconomic impact worldwide.

8.1 Pathogenesis of RA Arise from autoimmune inflammation and OS

The pathogenesis of RA involves A complex interplay between genetic and environmental factors leading to autoimmune inflammatory responses against the connective and synovial tissues of the joints [136]. Furthermore, increased ROS concentrations are actively involved in RA pathogenesis [206, 207]. Infiltration of the affected synovial tissues and promotion of the overexpression, release, and activity of proinflammatory cytokines [TNF-α, TNF-induced NF-κB, vascular endothelial growth factor (VEGF), IL-1 beta (IL-1β), IL-6, IL-8, and IFN-γ] by T cells, B cells, and macrophages are particular findings in patients with RA [208, 209].

Responding to the proinflammatory cytokines, synoviocytes (FLS) fibroblasts flourish and produce huge quantities of cytokines, matrix metalloproteinases (MMPs), and COX-2, which progressively degrade cartilage and lead to joint obliteration [210, 211]. Oxidative stress is involved in the disintegration of cartilage through NrF2 or NFE2L2 dysregulation [212]. Activated Nrf2 binds to antioxidant response elements (AREs), resulting in the augmented expression of antioxidative enzyme [e.g., Heme oxygenase-1 (HO-1)] encoding genes [213, 214] which indicates both OS and inflammatory response are implicated in the RA pathogenesis. Resultantly, phytochemicals with anti-inflammatory capacities play a crucial role in the battle with RA.

8.2 Phytochemicals against OS and inflammation in RA

Side effects are common and unavoidable from synthetic regimens used for managing RA making alternative medicine and traditional medicine a viable source for treating the disease. Phytochemicals attenuate OS and inflammation and relieve and protection from RA.

The clinical phenomenon in patients with RA involves osteoclastogenesis which is a process where bone tissues is destroyed by osteoclasts. **Polyphenols** extracted from **dried plums** extracted polyphenols inhibit osteoclastogenesis through suppressing TNF-α and NO synthase activity and by downregulating the transcription factor nuclear factor for activated T cells (NFATc1) [215]. **Cherry anthocyanins** reduce both OS (SOD and decrease serum MDA) and inflammatory mediators (decrease in TNF-α) in an SD adjuvant-induced RA rat model [216].

Resveratrol polyphenol confers significant protective effect against an aggressive RA rat model [217]. Anti-inflammatory and antioxidative resveratrol activities influenced reduction of specific rheumatoid biomarkers activities [serum rheumatoid factor (RF), MMP-3 and cartilage oligomeric matrix protein (COMP)], immunological biomarkers [IgG and antinuclear antibody (ANA)], immunomodulatory cytokines (TNF-α), and OS biomarkers [myeloperoxidase (MPO), CRP, and MDA] [217].

A natural polyphenol in mangoes, **mangiferin**, suppressed the expression of IL-1β, IL-6, TNF-α, and receptor activator of NF-κB ligand (RANKL) via the activation of extracellular signal-regulated kinase 1/2 (ERK1/2) and the inhibition of NF-κB [218]. Mangiferin also exerts strong proapoptotic effects on human synovia-derived synoviocytes when protecting against joint degeneration in RA [219].

Kaempferol from grapefruits inhibits synovial fibroblast proliferation by suppressing inflammatory cytokine IL-1β, inhibiting the phosphorylation of ERK-1/2, p38, and JNK, preventing the activation of NF-κB, and reducing OS by impeding the production of MMPs, COX-2, and PGE2 in RA-derived synovial fibroblasts [220].

9. Conclusion

Lipid peroxidation resulting from OS and inflammation are mixed up in the pathogeneses of degenerative brain disorders, DM, RA. Propositions targeting these provides means to develop viable strategies to treat these diseases using phytochemicals. Cells are furnished with antioxidant defense systems to combat the effects of OS with the Nrf2 being the master regulator of redox homeostasis. The regulator triggers the antioxidant enzyme systems. Consequently, targeting Nrf2 appears to offer a means of controlling OS. However, attenuating OS alone may not confer satisfactory protection against these diseases, in which case, targeting the classical cell survival pathway, that is, the TrkB/PI3K/Akt pathway would be required to restore cellular function. These signaling pathways upregulate pro-survival factors but suppress their pro-apoptotic counterparts.

Phytochemical with pharmacological modulation capacity may coactivate TrkB signaling mediated cell survival and Nrf2-ARE antioxidant systems. The combination offers promise for the treatment of diseases connected with OS-associated brain degeneration, glucose homeostasis derangements, and rheumatoid arthritis.

Contextually, several phytotherapeutics have been reported to protect against neuronal injury by activating TrkB/PI3K/Akt and Nrf2 signaling systems, which suggests they could be utilized to design novel therapeutic agents for NDD, ischemic stroke, TBI, and brain aging.

Phytochemicals (for example, resveratrol, tea polyphenols etc.) have been shown to promote the regeneration capacity of neurons along with their protection by dual targeting TrkB/PI3K and Nrf2-ARE signaling [81, 114]. These may have a better chance of succeeding with AD subjects.

Generally, the preventive and curative action of phytotherapeutics against pathological conditions [116, 221] emanate from their ability to behave as antioxidant and oxidants and that they are electron donors and electron receivers under varying environments in a pluripotential capacity which allows them to influence reduction and oxidation reactions [13, 32, 34, 35]. The negative values of redox potential probably enable the active principles to act as an antioxidant and in turn scavengers of free radicals as they are oxidized in the process. Crucially, it has been observed that the relative efficacy of antioxidant activity of the *Salix aegyptiaca* phytochemicals tends to be similar to their relative order of redox potential.

Acetylsalicylic acid has been shown to have lowest redox potential and antioxidant activity suggesting that the phytochemicals such as gallic acid, quercetin, rutin and vanillin, other than salicylates contribute to the medicinal properties of *S. aegyptiaca,* indicating a possible synergistic activity necessitating whole plant approaches in the use phytotherapeutics [222].

The interesting connection between OS, inflammation, lipid peroxidation are closely linked to initiation and progression of various diseases [223] necessitating interventions with phytochemicals to combat, at molecular level, different aspects of the biological homeostasis bringing about pathophysiological conditions of cardiovascular diseases, DM, RA and brain degenerative disorders of the old. The phytotherapeutics triterpenes Asiatic acid, maslinic and oleanolic have pleiotropic functions rendering them potent interventions for OS-related disease and lipid peroxidation.

Author details

Alfred Mavondo-Nyajena Mukuwa Greanious [1]*, Nesisa Ncube[2], Alfred Sibanda[1], Delton Dube[1], Francis Chikuse Farai[3] and Paul Makoni[4]

1 Pathology Department, Faculty of Medicine, National University of Science and Technology, Bulawayo, Zimbabwe

2 Biochemistry Department, Faculty of Medicine and Health Sciences, University of Zimbabwe, Harare, Zimbabwe

3 Regional COVID-19 Response and Vaccine Roll-out Coordinator, Windhoek, Namibia

4 Research and Internationalization Office, National University of Science and Technology, Bulawayo, Zimbabwe

*Address all correspondence to: greaniousa@gmail.com

References

[1] Rival T, Page R.M., Chandraratna D S, Sendall T J, Ryder E, and Liu B. (2009) Fenton chemistry and oxidative stress mediate the toxicity of the beta-amyloid peptide in a Drosophila model of Alzheimer's disease. Eur J Neurosci. 29(7):1335-1347

[2] Everett J, Cespedes E, Shelford L R, Exley C, Collingwood J F, and Dobson J. (2014) Ferrous iron formation following the co-aggregation of ferric iron and the Alzheimer's disease peptide beta-amyloid J R Soc Interface. 11(95): 20140165

[3] Lipinski B. (2012) Lipinski B. Is it oxidative or free radical stress and why does it matter Oxid Antioxid Med Sci. 1(1):1

[4] Birben E, Sahiner U, M., , Sackesen C, Erzurum S, and Kalayci O. (2012) Oxidative stress and antioxidant defense. World Allergy Organ J. 5(1):9-19

[5] Salvador G A, Uranga R M and Giusto N M. (2010) Iron and mechanisms of neurotoxicity. Int J Alzheimers Dis. 2011:720658

[6] Dorszewska J, Rożycka A, Oczkowska A, Florczak-Wyspiańska J, Prendecki M, and Dezor M. (2014) Mutations of TP53 Gene and Oxidative Stress in Alzheimer's Disease Patients. ADAPT. 3(1):24-32

[7] Dikalov S, Griendling K K and Harrison D G. (2007) Measurement of reactive oxygen species in cardiovascular studies. Hypertension. 49 (4):717-727

[8] Buetler T M, Krauskopf A and Ruegg U T. (2004) Role of superoxide as a signaling molecule. News Physiol Sci. 19:120-123

[9] Massey V. (1994) Activation of molecular oxygen by flavins and flavoproteins. J Biol Chem 269(36): 22459-22462

[10] Chaiyen P, Fraaije M W and Mattevi A. (2012) The enigmatic reaction of flavins with oxygen. . Trends Biochem Sci. 37(9):373-380

[11] Costa S, Reina-Couto M, Albino-Teixeira A, and Sousa T. (2016) Statins and oxidative stress in chronic heart failure. Revista Portuguesa de Cardiologia (English Edition). 35(1):41-57. https://doi.org/10.1016/j.repce.2015.12.016

[12] Tousoulis D, Charakida M, Stefanadi E, Siasos G, Latsios G, and Stefanadis C. (2007) Statins in heart failure. Beyond the lipid lowering effect. International Journal of Cardiology. 115(2):144-150.https://doi.org/10.1016/j.ijcard.2006.03.094

[13] Mavondo G A and Kasvosve I. (2017b) Antimalarial Phytochemicals: Delineation of the Triterpene Asiatic Acid Malarial Anti-Disease and Pathophysiological Remedial Activities-Part II. J Infect Disease Pathol 1:103

[14] Mavondo G A and Kasvosve I. (2017) Antimalarial Phytochemicals: Delineation of the Triterpene Asiatic Acid Malarial Anti-Disease and Pathophysiological Remedial Activities-Part I. J Infect Disease Pathol 1:104

[15] Niki E. (2014) Biomarkers of lipid peroxidation in clinical material. Biochimica et Biophysica Acta. 1840(2):809-817

[16] Esterbauer H, A. Benedetti, Lang J, Fulceri R, Fauler G, and Comporti M. (1986) Studies on the mechanism of formation of 4-hydroxynonenal during microsomal lipid peroxidation. Biochimica et Biophysica Acta: Lipids and Lipid Metabolism. 876(1):154-166

[17] Cheeseman K H, Beavis A and Esterbauer H. (1988) Hydroxylradical-induced iron-catalysed degradation of 2-deoxyribose. Quantitative determination of malondialdehyde. Biochemical J 252(3):649-653

[18] Esterbauer H and Zolliner H. (1989) Methods for determination of aldehydic lipid peroxidation products. Free Radical Biology and Medicine. 7(2):197-203

[19] Esterbauer H and Cheeseman K H. (1990) Determination of aldehydic lipid peroxidation products: malonaldehyde and 4-hydroxynonenal. Methods in Enzymology. 186:407-421

[20] Esterbauer H, Schaur R J and Zollner H. (1991) Chemistry and Biochemistry of 4-hydroxynonenal, malonaldehyde and related aldehydes. Free Radical Biology and Medicine 11(1):81-128

[21] Esterbauer H, Eckl P and Ortner A. (1990) Possible mutagens derived from lipids and lipid precursors. Mutation Research. 238:223-233

[22] Pryor W A. (1989) On the detection of lipid hydroperoxides in biological samples. Free Radical Biology and Medicine. 7(2):177-178

[23] Sinnhuber R O, Yu T C and Yu T C. (1958) Characterization of the red pigment formed in the 2-thiobarbituric acid determination of oxidative rancidity. J Food Science. 23(6):626-634

[24] Ayala A, Muñoz M F and Argüelles S. (2014) Lipid Peroxidation: Production, Metabolism, and Signaling Mechanisms of Malondialdehyde and 4-Hydroxy-2-Nonenal. Oxid Med Cell Longevity. 2014:360438.10.1155/2014/360438

[25] Schauenstein E. (1967) E. Schauenstein, "Autoxidation of polyunsaturated esters in water: chemical structure and biological activity of the products. J Lipid Research. 8(5):417-428

[26] A. Benedetti, Comporti M and Esterbauer H. (1980) Identification of 4-hydroxynonenal as a cytotoxic product originating from the peroxidation of liver microsomal lipids. Biochimica et Biophysica Acta: Lipids and Lipid Metabolism. 620(2):281-296

[27] Brambilla G, Sciab`a L and Faggin P. (1986) Cytotoxicity, DNA fragmentation and sister-chromatid exchange in Chinese hamster ovary cells exposed to the lipid peroxidation product 4-hydroxynonenal and homologous aldehydes. Mutation Research. 171(2-3):169-176

[28] Łuczaj W, Gęgotek A and Skrzydlewska E. (2017) Antioxidants and HNE in redox homeostasis. Free Radic Biol Med. 10.1016/j.freeradbiomed.2016.11.033

[29] Dalleau S, Baradat M, Guéraud F, and Huc L. (2013) Cell death and diseases related to oxidative stress: 4-hydroxynonenal (HNE) in the balance. Cell Death & Differentiation. 20(12):1615-1630.10.1038/cdd.2013.138

[30] Ray P D, Huang B-W and Tsuji Y. (2012) Reactive oxygen species (ROS) homeostasis and redox regulation in cellular signaling. Cellular Signalling. 24(5):981-990.10.1016/j.cellsig.2012.01.008

[31] Wu P S, Yen J H, Kou M C, and Wu M J. (2015) Luteolin and Apigenin Attenuate 4-Hydroxy-2-Nonenal-Mediated Cell Death through Modulation of UPR, Nrf2-ARE and MAPK Pathways in PC12 Cells. PLoS One. 10(6):e0130599.10.1371/journal.pone.0130599

[32] Greanious Alfred Mavondo, Joy Mavondo, Wisdom Peresuh, Dlodlo M, and Moyo O. Malaria Pathophysiology

as a Syndrome: Focus on Glucose Homeostasis in Severe Malaria and Phytotherapeutics Management of the Disease, in Parasites and Parasitic Diseases, Bastidas G, Editor. April 24th 2019, IntechOpen 24-52.

[33] Mavondo G A and Mzingwane M L. Severe Malarial Anemia (SMA) Pathophysiology and the Use of Phytotherapeutics as Treatment Options, in Current Topics in Anemia, Khan J, Editor. 2017.

[34] Mavondo G A, Mkhwanazi B N, Mzingwane M L, Dangarembizi R, Zambuko B, Moyo O, et al. Malarial Inflammation-Driven Pathophysiology and Its Attenuation by Triterpene Phytotherapeutics, in Parasitology and Microbiology Research, (eds) GABPaAAK, Editor. 2019, IntechOpen: London, SW7 2QJ, UNITED KINGDOM

[35] Mavondo G A, Mkhwanazi B N, Mavondo J, Peresuh W, and Moyo O. Malarial pathophysiology and phytochemical interventions: a current discourse on oxidative stress anti-disease phytotherapeutics, in Malaria. 2018: London.

[36] Mavondo G A, Mkhwanazi B M, Mhlanga C, Dangarembizi R, Mukonowenzou N, Mzingwane M L, et al. Oxidative Stress in Malarial Diseases: Plasmodium-Human Host Interactions and Therapeutic Interventions, in Oxidative Stress in Microbial Diseases, Sajal Chakraborti· Tapati Chakraborti, Dhrubajyoti Chattopadhyay, and (Editors) CS, Editors. 2019, Springer Nature Singapore 411-.

[37] Mkhwanazi B N, van Heerden F R, Mavondo G A, Mabandla M V, and Musabayane C T. (2019) Triterpene derivative improves the renal function of streptozotocin-induced diabetic rats: a follow-up study on maslinic acid. Renal Failure. 41(1):547-554. 10.1080/0886022X.2019.1623818

[38] Mkwananzi B N, Serumula M R, Myburg R B, van-Heerden F, and Musabayane C T. (2014) Antioxidant effects of maslinic acid in livers, hearts and kidneys of streptozotocin-induced diabetic rats: effects on kidney function. Renal Failure. 36:419-431

[39] Mkhwanazi B N, Serumula M R, Myburg R B, Van Heerden F, R., , and Musabayane C T. (2014) Antioxidant effects of maslinic acid in livers, hearts and kidneys of streptozotocin-induced diabetic rats: effects on kidney function. Renal Failure 36:419-431. 10.3109/0886022X.2013.867799

[40] Girotti A W. (1998) Lipid hydroperoxide generation, turnover, and effector action in biological systems. J Lipid Research. 39(8): 1529-1542

[41] Zoidis E, Seremelis I, Kontopoulos N, and Danezis G P. (2018) Selenium-Dependent Antioxidant Enzymes: Actions and Properties of Selenoproteins. Antioxidants (Basel). 7(5):66.10.3390/antiox7050066

[42] Brigelius-Flohé R and Maiorino M. (2013) Glutathione peroxidases. Biochimica et Biophysica Acta. 1830(5):3289-3303

[43] Asgarpanah J. (2012) Phytopharmacology and medicinal properties of Salix aegyptiaca L. Afr J Biotechnol. 11 (28):7145-7750

[44] Karimi I, Hayatgheybi H, Kamalak A, Pooyanmehr M, and Marandi Y. (2011) Chemical composition and effect of an essential oil of Salix aegyptiaca L., Salicaceae, (musk willow) in hypercholesterolemic rabbit model. Rev Bras. 21(3):407-414

[45] Mahdi J, Mahdi A and Bowen I. (2006) The historical analysis of aspirin discovery, its relation to the willow tree and antiproliferative and anticancer

potential. Cell Prolif. 2006;. 39 (2):147-155

[46] Enaya t S and Banerjee S. (2009) Comparative antioxidant activity of extracts from leaves, bark and catkins of *Salix aegyptiaca* sp. Food Chem 116(1)::23-28

[47] Christen Y. (2000) Oxidative stress and Alzheimer disease. Am J Clin Nutr. 71(2):621s-629s

[48] Wu L L, Chiou C C, Chang P Y, and Wu J T. (2004) Urinary 8-OHdG: a marker of oxidative stress to DNA and a risk factor for cancer, atherosclerosis and diabetics. Clin Chim Acta. 339(1):1-9

[49] Chen M-S, Xu F-P, Wang Y-Z, Zhang G-P, Yi Q, Zhang H-Q, et al. (2004) Statins initiated after hypertrophy inhibit oxidative stress and prevent heart failure in rats with aortic stenosis. Journal of Molecular and Cellular Cardiology. 37(4):889-896.https://doi.org/10.1016/j.yjmcc.2004.06.019

[50] Nauman M, Kale R K and Singh R P. (2018) Polyphenols of Salix aegyptiaca modulate the activities of drug metabolizing and antioxidant enzymes, and level of lipid peroxidation. BMC Complementary and Alternative Medicine 19:81.https://doi.org/10.1186/s12906-018-2143-7

[51] CHENG A, CHEN X, 1, JIN Q, WANG W, SHI J, and LIU Y. (2013) Comparison of Phenolic Content and Antioxidant Capacity of Red and Yellow Onions. Czech J. Food Sci. 31(5):501-508

[52] Sharma S, Arif M, Nirala R K, Gupta R, and Thakur S C. (2015) Cumulative therapeutic effects of phytochemicals in Arnica montana flower extract alleviated collagen-induced arthritis: inhibition of both pro-inflammatory mediators and oxidative stress. J Sci Food Agric 10.1002/jsfa.7252

[53] Winterbourn C C, Buss H I, Chan T P, Plank L D, Clark M A, and Windsor J A. (2000) Protein carbonyl measurements show evidence of early oxidative stress in critically ill patients. Critical Care Medicine. 28(1)

[54] Matsuyama Y, Terawaki H, Terada T, and Era S. (2009) Albumin thiol oxidation and serum protein carbonyl formation are progressively enhanced with advancing stages of chronic kidney disease. Clinical and Experimental Nephrology. 13(4):308-315.10.1007/s10157-009-0161-y

[55] Song Y R, Kim J-K, Lee H-S, Kim S G, and Choi E-K. (2020) Serum levels of protein carbonyl, a marker of oxidative stress, are associated with overhydration, sarcopenia and mortality in hemodialysis patients. BMC Nephrology. 21(1):281.10.1186/s12882-020-01937-z

[56] Colombo G, Reggiani F, Angelini C, Finazzi S, Astori E, Garavaglia M L, et al. (2020) Plasma Protein Carbonyls as Biomarkers of Oxidative Stress in Chronic Kidney Disease, Dialysis, and Transplantation. Oxidative Medicine and Cellular Longevity. 2020:2975256.10.1155/2020/2975256

[57] Adefegha S A and Oboh G. (2011) Water Extractable Phytochemicals from Some Nigerian Spices Inhibit Fe2+-Induced Lipid Peroxidation in Rat's Brain-In Vitro J Food Process Technol 2:104.10.4172/2157-7110.1000104

[58] Kumar S, Sharma U K, Sharma A, and Pandey A K. (2012) Protective Efficacy of Solanum *xanthocarpum* root extracts against free radical damage: phytochemical analysis analysis and antioxidant effect. Cell Mol Biol. 58(1):174-181.10.1170/T938

[59] Mansour M A, Nagi M N, El-Khatib A S, and Al-Bekairi A M. (2002) Effects of thymoquinone on antioxidant enzyme activities, lipid peroxidation

and DT-diaphorase in different tissues of mice: a possible mechanism of action. Cell Biochemistry and Function. 20(2):143-151.https://doi.org/10.1002/cbf.968

[60] Sharma R A, Ireson C R, Verschoyle R D, Hill K A, Williams M L, Leuratti C, et al. (2001) Effects of Dietary Curcumin on Glutathione S-Transferase and Malondialdehyde-DNA Adducts in Rat Liver and Colon Mucosa. Clin Cancer Res 7(5):1452-1458

[61] Ighodaro O M and Akinloye O A. (2018) First line defence antioxidants-superoxide dismutase (SOD), catalase (CAT) and glutathione peroxidase (GPX): Their fundamental role in the entire antioxidant defence grid. Alexandria Journal of Medicine. 54(4):287-293.https://doi.org/10.1016/j.ajme.2017.09.001

[62] Hannan A, Dash R, Sohag A A, Haque N, and Moon I S. (2020a) Neuroprotection Against Oxidative Stress: Phytochemicals Targeting TrkB Signaling and the Nrf2-ARE Antioxidant System. Front Mol Neurosci. 13:116.10.3389/fnmol.2020.00116

[63] Sivandzade F, Prasad S, Bhalerao A, and Cucullo L. (2019) NRF2 and NF-kB interplay in cerebrovascular and neurodegenerative disorders: molecular mechanisms and possible therapeutic approaches. Redox Biol. 21:101059

[64] Niedzielska E, Smaga I, Gawlik M, Moniczewski A, Stankowicz P, and Pera J. (2016) Oxidative stress in neurodegenerative diseases. Mol Neurobiol. 53:4094-4125

[65] Rodrigo R, Fernández-Gajardo R, Gutiérrez R, Matamala J M, Carrasco R, and Miranda-Merchak A. (2013) Oxidative stress and pathophysiology of ischemic stroke: novel therapeutic opportunities. CNS Neurol Disord Drug Targets 12:698-714.10.2174/18715273113 12050015

[66] Rodríguez-Rodríguez A, Egea-Guerrero J J, Murillo-Cabezas F, and Carrillo-Vico A. (2014) Oxidative stress in traumatic brain injury. Curr Med Chem. 21:1201-1211

[67] Liu Z, Zhou T, Ziegler A C, Dimitrion P, and Zuo L. (2017) Oxidative stress in neurodegenerative diseases: from molecular mechanisms to clinical applications. Oxid Med Cell Longev. 2017:2525967-2525967

[68] Wang X, Wang W, Li L, Perry G, Lee H-G, and Zhu X. (2014) Oxidative stress and mitochondrial dysfunction in Alzheimer's disease. Biochim Biophys Acta 1842:1240-1247

[69] Sampaio T, Savall A, Gutierrez M, and Pinton S. (2017) Neurotrophic factors in Alzheimer's and Parkinson's diseases: implications for pathogenesis and therapy. Neural Regen Res 12:549-557

[70] Gupta V K, You Y, Gupta V B, Klistorner A, and Graham S L. (2013) TrkB receptor signalling: implications in neurodegenerative, psychiatric and proliferative disorders. Int J Mol Sci 14:10122-10142.10.3390/ijms140510122

[71] Mitre M, Mariga A and Chao M V. (2017) Neurotrophin signalling: novel insights into mechanisms and pathophysiology. Clin Sci (London, England 1979) 131:13-23.10.1042/cs20160044

[72] Amato A, Terzo S and Mulè F. (2019) Natural compounds as beneficial antioxidant agents in neurodegenerative disorders: a focus on Alzheimer's disease. Antioxidants Basel, Switzerland 8:608.10.3390/antiox8120608

[73] Lee J, Jo D-G, Park D, Chung H Y, and Mattson M P. (2014) Adaptive cellular stress pathways as therapeutic

targets of dietary phytochemicals: focus on the nervous system. Pharmacol Rev. 66:815-868.10.1124/pr.113.007757

[74] Hannan M A, Sohag A A M, Dash R, Haque M N, Mohibbullah M, and Oktaviani D F. (2020c) Phytosterols of marine algae: insights into the potential health benefits and molecular pharmacology Phytomedicine 691:53201.10.1016/j.phymed.2020. 15320

[75] Hannan M A, Dash R, Sohag A A M, and Moon I S. (2019) Deciphering molecular mechanism of the neuropharmacological action of fucosterol through integrated system pharmacology and in silico analysis. Mar Drugs. 17:639.10.3390/md17110639

[76] Hannan M A, Kang J Y, Mohibbullah M, Hong Y K, Lee H, C, and hoi J S. (2015) *Moringa oleifera* with promising neuronal survival and neurite outgrowth promoting potentials. J Ethnopharmacol. 152:142-150.10. 1016/j.jep.2013.12.036

[77] Moosavi F, Hosseini R, Saso L, and Firuzi O. (2015) Modulation of neurotrophic signaling pathways by polyphenols. Drug Des Dev Ther. 10:23-42

[78] Tohda C. (2016) New age therapy for Alzheimer's disease by neuronal network reconstruction. Biol Pharm Bull 39:1569-1575

[79] Gao J, Liu S, Xu F, Liu Y, Lv C, and Deng Y. (2018) Trilobatin protects against oxidative injury in neuronal PC12 cells through regulating mitochondrial ROS homeostasis mediated by AMPK/Nrf2/Sirt3 signaling pathway. Front Mol Neurosci 11:267.10.3389/fnmol.2018.00267

[80] Cui H-Y, Zhang X-J, Yang Y, Zhang C, Zhu C-H, and Miao J-Y. (2018) Rosmarinic acid elicits neuroprotection in ischemic stroke via Nrf2 and heme oxygenase 1 signaling. Neural Regen Res 13:2119-2128

[81] Hui Y, Chengyong T, Cheng L, Haixia H, Yuanda Z, and Weihua Y. (2018) Resveratrol attenuates the cytotoxicity induced by amyloid-b(1-42) in PC12 cells by upregulating heme oxygenase-1 via the PI3K/Akt/Nrf2 pathway. Neurochem Res 43:297-305.10.1007/s11064-017-2421-7

[82] Gandhi S and Abramov A Y. (2012) Mechanism of oxidative stress in neurodegeneration. Oxid Med Cell Longev. 2012:428010

[83] Galasko D R, Peskind E, Clark C M, Quinn J F, Ringman J M, and Jicha G A. (2012) Antioxidants for Alzheimer disease: a randomized clinical trial with cerebrospinal fluid biomarker measures. Arch Neurol 69:836-841

[84] Butterfield D A and Boyd-Kimball D. (2018) Oxidative stress, amyloid-b peptide, and altered key molecular pathways in the pathogenesis and progression of Alzheimer's disease. J Alzheimers Dis. 62:1345-1367.10.3233/jad-170543

[85] Butterfield D A and Boyd-Kimball D. (2019) Redox proteomics and amyloid b-peptide: insights into Alzheimer disease. J Neurochem. 151:459-487.10.1111/ jnc.14589

[86] van der Kant R, Goldstein L S B and Ossenkoppele R. (2020) Amyloid-B-independent regulators of tau pathology in Alzheimer disease. Nat Rev Neurosci. 21:21-35

[87] Wojsiat J, Zoltowska K M, Laskowska-Kaszub K, and Wojda U. (2018) Oxidant/antioxidant imbalance in Alzheimer's disease: therapeutic and diagnostic prospects. Oxid Med Cell Longev. 2018:6435861

[88] Youssef P, Chami B, Lim J, Middleton T, Sutherland G T, and Witting P K. (2018) Evidence supporting oxidative stress in a moderately affected area of the brain in Alzheimer's disease. Sci Rep. 11553:11553

[89] Di Domenico F, Tramutola A and Butterfield D A. (2017) Role of 4-hydroxy-2-nonenal (HNE) in the pathogenesis of Alzheimer disease and other selected age-related neurodegenerative disorders. Free Radic Biol Med 111:253-261.10.1016/j.freeradbiomed.2016.10.490

[90] Sharma A, Weber D, Raupbach J, Dakal T C, Fließbach K, and Ramirez A. (2020) Advanced glycation end products and protein carbonyl levels in plasma reveal sex-specific differences in Parkinson's and Alzheimer's disease. Redox Biol. 34:101546

[91] Tramutola A, Abate G, Lanzillotta C, Triani F, Barone E, and Iavarone F. (2018) Protein nitration profile of CD3(+) lymphocytes from Alzheimer disease patients: novel hints on immunosenescence and biomarker detection. Free Radic Biol Med 129:430-439

[92] Kim T-S, Pae C-U, Yoon S-J, Jang W-Y, Lee N J, and Kim J-J. (2006) Decreased plasma antioxidants in patients with Alzheimer's disease. Int J Geriatr Psychiatry 21:344-348

[93] Singh A, Kukreti R, Saso L, and, and Kukreti S. (2019) Oxidative stress: a key modulator in neurodegenerative diseases. Molecules (Basel, Switzerland). 24:1583

[94] Buendia I, Michalska P, Navarro E, Gameiro I, Egea J, and León R. (2016) Nrf2-ARE pathway: an emerging target against oxidative stress and neuroinflammation in neuro degenerative diseases. Pharmacol Ther. 157:84-104.10.1016/j.pharmthera.2015.11.003

[95] Tönnies E and Trushina E. (2017) Oxidative stress, synaptic dysfunction, and Alzheimer's disease. J Alzheimers Dis. 57:1105-1121

[96] Gaki G S and Papavassiliou A G. (2014) Oxidative stress-induced signaling pathways implicated in the pathogenesis of Parkinson's disease. Neuro Mol Med 16:217-230.10.1007/s12017-014-8294-x

[97] de Farias C C, Maes M, Bonifácio K L, Bortolasci C C, De Souza Nogueira, and A. B, F. F. . (2016) Highly specific changes in antioxidant levels and lipid peroxidation in Parkinson's disease and its progression: disease and staging biomarkers and new drug targets. Neurosci Lett. 617:66-71.10.1016/j.neulet.2016.02.011

[98] Beal M F. (2002) Oxidatively modified proteins in aging and disease. Free Radic Biol Med. 32:797-803.10.1016/s0891-5849(02)00780-3

[99] Wei Z, Li X, Li X, Liu Q, and Cheng Y. (2018) Oxidative stress in Parkinson's disease: a systematic review and meta-analysis. Front Mol Neurosci 11:236.10.3389/fnmol.2018.00236

[100] Donkor E S. (2018) Stroke in the 21(st) century: a snapshot of the burden, epidemiology, and quality of life. Stroke Res Treat 2018:3238165

[101] Feigin V L, Norrving B and Mensah G A. (2017) Global burden of stroke. Circ Res. 120:439-448

[102] Hannan M A, Haque M N, Mohibbullah M, Dash R, Hong Y K, and Moon I S. (2020b) Gelidium amansii attenuates hypoxia/reoxygenation-induced oxidative injury in primary hippocampal neurons through suppressing GluN2B expression.

Antioxidants (Basel) 9:223.10.3390/antiox9030223

[103] Wu C-H, Chen C-C, Hung T-H, Chuang Y-C, Chao M, and Shyue S-K. (2019) Activation of TrkB/Akt signaling by a TrkB receptor agonist improves long-term histological and functional outcomes in experimental intracerebral hemorrhage. J Biomed Sci 26:53

[104] Kishimoto M, Suenaga J, Takase H, Araki K, Yao T, and Fujimura T. (2019) Oxidative stress-responsive apoptosis inducing protein (ORAIP) plays a critical role in cerebral ischemia/reperfusion injury. Sci Rep 9:13512

[105] Dogan O, Kisa U, Erdemoglu Ali K, Kacmaz M, Caglayan O, and Kurku H. (2018) Oxidative and nitrosative stress in patients with ischemic stroke. Laboratoriums Medizin 42:195-200.10.1515/labmed-2018-0036

[106] Lozano D, Gonzales-Portillo G S, Acosta S, De La Pena I, and Tajiri N, Kaneko, Y. (2015) Neuroinflammatory responses to traumatic brain injury: etiology, clinical consequences, and therapeutic opportunities. Neuropsychiatr Dis Treat 11:97-106

[107] Khatri N, Thakur M, Pareek V, Kumar S, Sharma S, and Datusalia A K. (2018) Oxidative stress: major threat in traumatic brain injury. CNS Neurol Disord Drug Targets 17:689-695.10.2174/1871527317666180627120501

[108] Wurzelmann M, Romeika J and Sun D. (2017) Therapeutic potential of brain-derived neurotrophic factor (BDNF) and a small molecular mimics of BDNF for traumatic brain injury. Neural Regen Res 12:7-12

[109] Houlton J, Abumaria N, Hinkley S F R, and Clarkson A N. (2019) Therapeutic potential of neurotrophins for repair after brain injury: a helping hand from biomaterials. Front Neurosci 13: 790.10.3389/fnins.2019.00790

[110] Sun Y, Yang T, Leak R K, Chen J, and Zhang F. (2017) Preventive and protective roles of dietary Nrf2 activators against central nervous system diseases. CNS Neurol Disord. Drug Targets 16:326-338

[111] Kwon S H, Ma S X, Hwang J Y, Lee S Y, and Jang C G. (2015) Involvement of the Nrf2/HO-1 signaling pathway in sulfuretin-induced protection against amyloid beta 25-35 neurotoxicity. Neuroscience 304:14-28.10.1016/j.neuroscience.2015.07.030

[112] Kim S, Song N-J, Chang S-H, Bahn G, Choi Y, Rhee D-K, et al. (2019) Sulfuretin Prevents Obesity and Metabolic Diseases in Diet Induced Obese Mice Biomolecules & Therapeutics 27(1):107-116 https://doi.org/10.4062/biomolther.2018.090

[113] Ali T, Kim T, Rehman S U, Khan M S, Amin F U, and Khan M. (2018) Natural dietary supplementation of anthocyanins via PI3K/Akt/Nrf2/HO-1 pathways mitigate oxidative stress, neurodegeneration, and memory impairment in a mouse model of Alzheimer's disease. Mol. Neurobiol. 55:6076-6093.10.1007/s12035-017-0798-6

[114] Qi G, Mi Y, Wang Y, Li R, Huang S, and Li X. (2017) Neuroprotective action of tea polyphenols on oxidative stress-induced apoptosis through the activation of the TrkB/CREB/BDNF pathway and Keap1/Nrf2 signaling pathway in SH-SY5Y cells and mice brain. Food Funct. 8:4421-4432. 10.1039/c7fo00991g

[115] Wu P-S, Ding H-Y, Yen J-H, Chen S-F, Lee K-H, and Wu M-J. (2018) Anti-inflammatory activity of 8-hydroxydaidzein in LPS-stimulated BV2 microglial cells via activation of Nrf2-antioxidant and attenuation of Akt/NF-kB-inflammatory signaling pathways, as well as inhibition of COX-2 activiy J Agric Food Chem 66:5790-5801

[116] Silva R R, Silva D O, Fontes H R, Alviano C S, Fernandes P D, and Alviano D S. (2013) Antiinflammatory, antioxidant, and antimicrobial activities of Cocos nucifera var. typica. BMC Complement Altern Med. 13:107

[117] Tranchimand S, Brouant P and Iacazio G. (2010) The rutin catabolic pathway with special emphasis on quercetinase. Biodegradation. 21(6):833-859.10.1007/s10532-010-9359-7

[118] Thabet N M and Moustafa E M. (2018) Protective effect of rutin against brain injury induced by acrylamide or gamma radiation: role of PI3K/AKT/GSK-3b/NRF-2 signalling pathway Arch Physiol Biochem. 124:185-193

[119] Enogieru A B, Haylett W, Hiss D C, Bardien S, and Ekpo O E. (2018) Rutin as a Potent Antioxidant: Implications for Neurodegenerative Disorders. Oxidative Medicine and Cellular Longevity. 2018:6 241017.10.1155/2018/6241017

[120] Budzynska B, Faggio C, Kruk-Slomka M, Samec D, Nabavi S F, Sureda A, et al. (2019) Rutin as Neuroprotective Agent: From Bench to Bedside. Curr Med Chem. 26(27):5152-5164.10.2174/092986732466617 1003114154

[121] Hao G, Dong Y, Huo R, Wen K, Zhang Y, and Liang G. (2016) Rutin Inhibits Neuroinflammation and Provides Neuroprotection in an Experimental Rat Model of Subarachnoid Hemorrhage, Possibly Through Suppressing the RAGE-NF-κB Inflammatory Signaling Pathway. Neurochem Res 41(6):1496-1504.10.1007/s11064-016-1863-7

[122] Song H L, Zhang X, Wang W Z, Liu R H, Zhao K, Liu M Y, et al. (2018) Neuroprotective mechanisms of rutin for spinal cord injury through anti-oxidation and anti-inflammation and inhibition of p38 mitogen activated protein kinase pathway. Neural Regen Res. 13(1):128-134.10.4103/1673-5374.217349

[123] Lee H, Ko, W., , Chowdhury A, Li B, Kim S C, and Oh H. (2020) Brassicaphenanthrene A from Brassica rapa protects HT22 neuronal cells through the regulation of Nrf2-mediated heme oxygenase-1 expression. Mol Med Rep 21:493-500

[124] Lee D-S, Cha B-Y, Woo J-T, Kim Y-C, and Jang J-H. (2015) Acerogenin a from acer nikoense maxim prevents oxidative stress-induced neuronal cell death through Nrf2-mediated heme oxygenase-1 expression in mouse hippocampal HT22 cell line. Molecules (Basel, Switzerland) 20:12545-12557.10.3390/molecules200712545

[125] Byun E-B, Cho E-J, Kim Y-E, Kim W S, and Byun E-H. (2018) Neuroprotective effect of polysaccharide separated from Perilla *frutescens* Britton var. acuta Kudo against H(2)O(2)-induced oxidative stress in HT22 hippocampus cells. Biosci Biotechnol Biochem 82:1344-1358.10.1080/ 09168451.2018.1460572

[126] Lee B D, Yoo J-M, Baek S Y, Li F Y, Sok D-E, and Kim M R. (2019)3,3'-Diindolylmethane promotes BDNF and antioxidant enzyme formation via TrkB/Akt pathway activation for neuroprotection against oxidative stressinduced apoptosis in hippocampal neuronal cells. Antioxidants 9:3.10.3390/antiox9010003

[127] Xu X H, Li G L, Wang B A, Qin Y, Bai S R, and Rong J. (2015) Diallyl trisufide protects against oxygen glucose deprivation -induced apoptosis by scavenging free radicals via the PI3K/Akt -mediated Nrf2/HO-1 signaling pathway in B35 neural cells. Brain Res. 1614:38-50

[128] Ge X-H, Shao L and Zhu G-J. (2018) Oxymatrine attenuates brain

hypoxic-ischemic injury from apoptosis and oxidativep-Akt/GSK3b/HO-1/Nrf-2 signaling pathway. Metabol Brain Di. 33:1869-1875.10.1007/s11011-018-0293-4

[129] Wen Z, Hou W, Wu W, Zhao Y, Dong X, and Bai X. (2018)6'-O-Galloylpaeoniflorin attenuates cerebral ischemia reperfusion-induced neuro inflammation and oxidative stress via PI3K/Akt/Nrf2 activation. Oxid Med Cell Longev 2018:8678267

[130] Zhang W, Song J-K, Yan R, Li L, Xiao Z-Y, and Zhou W-X. (2018) Diterpene ginkgolides protect against cerebral ischemia/reperfusion damage in rats by activating Nrf2 and CREB through PI3K/Akt signaling. Acta Pharmacol Sin 39:1259-1272

[131] Liu Q, Jin Z, Xu Z, Yang H, Li L, and Li G. (2019) Antioxidant effects of ginkgolides and bilobalide against cerebral ischemia injury by activating the Akt/Nrf2 pathway in vitro and in vivo. Cell Stress Chaperones 24:441-452.10.1007/s12192-019-00977-1

[132] Shu K and Zhang Y. (2019) Protodioscin protects PC12 cells against oxygen and glucose deprivation-induced injury through miR-124/AKT/Nrf2 pathway. Cell Stress Chaperones 24:1091-1099

[133] Liu X, Zhang X, Ma K, Zhang R, Hou P, and Sun B. (2016) Matrine alleviates early brain injury after experimental subarachnoid hemorrhage in rats: possible involvement of PI3K/Akt-mediated NF-kB inhibition and Keap1/Nrf2-dependent HO-1 induction. Cell Mol Biol (Noisy-le-Grand, France) 62:38-44

[134] Hu S, Liu T, Wu Y, Y, ang W, Hu S, and Sun Z. (2019) Panax notoginseng saponins suppress lipopolysaccharide-induced barrier disruption and monocyte adhesion on bEnd.3 cells via the opposite modulation of Nrf2 antioxidant and NF-kB inflammatory pathways. Phytother Res 33:3163-3176.10.1002/ptr.6488

[135] Islam A, Alam F, Solayman, Khalil I, Kamal M K, and Gan S H. (2016) Dietary Phytochemicals: Natural Swords Combating Inflammation and Oxidation-Mediated Degenerative Diseases. Hindawi Publishing Corporation Oxidative Medicine and Cellular Longevity. 2016 (Article ID 5137431):25 pages.http://dx.doi.org/10.1155/2016/5137431

[136] McInnes I B and Schett G. (2011) The pathogenesis of rheumatoid arthritis. The New England J Medicine. 365:2205-2219

[137] Odegaard A O, Jacobs D R, Sanchez O A, D. C. Goff D C, Reiner A P, and Gross M D. (2016) Oxidative stress, inflammation, endothelial dysfunction and incidence of type 2 diabetes. Cardiovascular Diabetology. 15(1):1-12

[138] L¨uscher T F L. (2015) Ageing, inflammation, and oxidative stress: final common pathways of cardiovascular disease. European Heart J. 36(48):3381-3383

[139] Alagona P and Ahmad T A. (2015) Cardiovascular disease risk assessment and prevention: current guidelines and limitations. Medical Clinics of North America. 99(4):711-731

[140] Kokil G R, Veedu R N, Ramm G A, Prins J B, and Parekh H S. (2015) Type 2 diabetes mellitus: limitations of conventional therapies and intervention with nucleic acid-based therapeutics. Chemical Reviews. 115(11):4719-4743

[141] Albrecht K and M¨uller-Ladner U. (2010) Side effects and management of side effects of methotrexate in rheumatoid arthritis. Clinical and Experimental Rheumatology. 28(5 supplement):S95-S101

[142] Franks P W, Hanson R L, Knowler W C, Sievers M L, Bennett P H, and Looker H C. (2010) Childhood obesity, other cardiovascular risk factors, and premature death. The New England Journal of Medicine. 362(6):485-493

[143] Singh R B, Mengi S A, Xu Y-J, Arneja A S, and Dhalla N S. (2002) Pathogenesis of atherosclerosis: a multifactorial process. Experimental and Clinical Cardiology. 7(1):40-53

[144] Goncharov N V, Avdonin P V, Nadeev A D, Zharkikh I L, and Jenkins R O. (2015) Reactive oxygen species in pathogenesis of atherosclerosis. Current Pharmaceutical Design. 21(9):1134-1146

[145] Takachi R, Inoue M and Ishihara J. (2008) Fruit and vegetable intake and risk of total cancer and cardiovascular disease Japan Public Health Center-based Prospective study. American Journal of Epidemiology. 167(1):59-70

[146] Wang X, Ouyang Y and Liu J. (2014) Fruit and vegetable consumption and mortality from all causes, cardiovascular disease, and cancer: systematic review and dose-response meta-analysis of prospective cohort studies. British Medical J. 349(Article ID g4490):1-14

[147] X. Gao X, Bermudez O I and Tucker K L. (2004) Plasma C-reactive protein and homocysteine concentrations are related to frequent fruit and vegetable intake in Hispanic and non-Hispanic white elders. J Nutrition. 134(4):913-918

[148] Du H, Li L and Bennett D. (2016) Fresh fruit consumption and major cardiovascular disease in China. The New England J Medicine. 374(14): 1332-1343

[149] Nagasako-Akazome Y N, Kanda T, Ohtake Y, Shimasaki H, and Kobayashi T. (2007) Apple polyphenols influence cholesterol metabolism in healthy subjects with relatively high body mass index. J Oleo Science. 56(8):417-428

[150] Yin X, Quan J and Kanazawa T. (2008) Banana prevents plasma oxidative stress in healthy individuals. Plant Foods for Human Nutrition. 63(2):71-76

[151] Basu A, Du M and Leyva M J. (2010a) Blueberries decrease cardiovascular risk factors in obese men and women with metabolic syndrome. The Journal of Nutrition. 140(9):1582-1587

[152] Basu A, Rhone M and Lyons T J. (2010b)Berries: emerging impact on cardiovascular health. Nutrition Reviews. 68(3):168-177

[153] Toh J Y, Tan V M H, Lim P C Y, Lim S T, and Chong M F F. (2013) Flavonoids fromfruit and vegetables: a focus on cardiovascular risk factors. Current Atherosclerosis Reports. 15(12):1-7

[154] Morand C, Dubray C and Milenkovic D. (2011) Hesperidin contributes to the vascular protective effects of orange juice: a randomized crossover study in healthy volunteers. American J Clinical Nutrition. 93(1):73-80

[155] Rizza S, Muniyappa R and Iantorno M. (2011) Citrus polyphenol hesperidin stimulates production of nitric oxide in endothelial cells while improving endothelial function and reducing inflammatory markers in patients with metabolic syndrome. The J Clinical Endocrinology & Metabolism. 96(5):E782-E792

[156] Ismail T, Sestili P and Akhtar S. (2012) Pomegranate peel and fruit extracts: a review of potential anti-inflammatory and antiinfective effects. J Ethnopharmacology. 143(2):397-405

[157] Noratto G, Martino H S D, Simbo S, Byrne D, and Mertens-Talcott

S U. (2015) Consumption of polyphenol-rich peach and plum juice prevents risk factors for obesity-related metabolic disorders and cardiovascular disease in Zucker rats. J Nutritional Biochemistry. 26(6):633-641

[158] Hong M Y, Hartig N, Kaufman K, Hooshmand S, Figueroa A, and Kern M. (2015) Watermelon consumption improves inflammation and antioxidant capacity in rats fed an atherogenic diet. Nutrition Research. 35(3):251-258

[159] Vazquez-Prieto M A and Miatello R M. (2010) Organosulfur compounds and cardiovascular disease. Molecular Aspects of Medicine. 31(6):540-545

[160] Chang H S, Yamato O, Yamasaki M, and Maede Y. (2005) Modulatory influence of sodium 2-propenyl thiosulfate from garlic on cyclooxygenase activity in canine platelets: possiblemechanism for the anti-aggregatory effect. Prostaglandins Leukotrienes and Essential Fatty Acids. 72(5):351-355

[161] Terao J, Kawai Y and Murota K. (2008) Vegetable flavonoids and cardiovascular disease. Asia Pacific J Clinical Nutrition. 17(supplement 1):291-293

[162] Upaganlawar A, Patel V and Balaraman R. (2012) Tomato lycopene attenuates myocardial infarction induced by isoproterenol: electrocardiographic, biochemical and anti-apoptotic study. Asian Pacific J Tropical Biomedicine. 2(5):345-351

[163] Pereira B L B, Arruda F C O and Reis P P. (2015) Tomato (*Lycopersicon esculentum*) supplementation induces changes in cardiac miRNA expression, reduces oxidative stress and left ventricular mass, and improves diastolic function,. Nutrients. 7(11):9640-9649

[164] Grzanna R, Lindmark L and Frondoza C G. (2005) Ginger-an herbal medicinal product with broad anti-inflammatory actions. J Medicinal Food. 8(2):125-132

[165] Ghayur M N and Gilani A H. (2005) Ginger lowers blood pressure through blockade of voltage-dependent calcium channels. J Cardiovascular Pharmacology. 45(1):74-80

[166] Fuhrman B, Rosenblat M, Hayek T, Coleman R, and Aviram M. (2000) Ginger extract consumption reduces plasma cholesterol, inhibits LDL oxidation and attenuates development of atherosclerosis in atherosclerotic, apolipoprotein E-deficient mice. J Nutrition. 130(5):1124-1131

[167] Wakade A S, Shah A S, Kulkarni M P, and A. R. Juvekar A R. (2008) Protective effect of Piper longum L. on oxidative stress induced injury and cellular abnormality in adriamycin induced cardiotoxicity in rats. Indian J Experimental Biology,. 46(7):528-533

[168] Vijayakumar R S, Surya D, Senthilkumar R, and Nalini N. (2002) Hypolipidemic effect of black pepper (Piper nigrum Linn.) in Rats fed high fat diet. J Clinical Biochemistry and Nutrition. 32:31-42

[169] Pickup J C. (2004) Inflammation and activated innate immunity in the pathogenesis of type 2 diabetes. Diabetes Care. 27(3):813-823

[170] Shoelson S E, Lee J and Goldfine A B. (2006) Inflammation and insulin resistance. The J Clinical Investigation. 116(7):1793-1801

[171] Werner E D, Lee J, Hansen L, Yuan M, and Shoelson S E. (2004) Insulin resistance due to phosphorylation of insulin receptor substrate-1 at serine 302. The J Biological Chemistry. 279(34):35298-35305

[172] Zick Y. (2005) Ser/Thr phosphorylation of IRS proteins: a

molecular basis for insulin resistance. Science Signaling. 2005(268): article pe4

[173] Alam F, Islam M A, Khalil M I, and Gan S H. (2016) Metabolic control of type 2 diabetes by targeting the GLUT4 glucose transporter: intervention approaches. Current Pharmaceutical Design. 22(20):3034-3049

[174] Cai D, Yuan M and Frantz D F. (2005) Local and systemic insulin resistance resulting from hepatic activation of IKK-B and NF-kB. Nature Medicine. 11(2):183-190

[175] Asmat U, Abad K and Ismail K. (2015) Diabetes mellitus and oxidative stress-a concise review. Saudi Pharmaceutical J. 2015

[176] Solayman M, Ali Y and Alam F. (2016) Polyphenols: potential future arsenals in the treatment of diabetes. Current Pharmaceutical Design. 22 (5):549-565

[177] Murunga A N, D. O. Miruka D O, Driver C, Nkomo F S, Cobongela S Z, and Owira P M. (2016) Grapefruit derived flavonoid naringin improves ketoacidosis and lipid peroxidation in type 1 diabetes rat model. PLoSONE. 11(4):1-16

[178] Sharma A K, Bharti S and Ojha S. (2011) Up-regulation of PPAR Gamma, heat shock protein-27 and-72 by naringin attenuates insulin resistance, B-cell dysfunction, hepatic steatosis and kidney damage in a rat model of type 2 diabetes. British J Nutrition. 106(11): 1713-1723

[179] Sekhon-Loodu S S, Ziaullah and Rupasinghe H P V. (2015) Docosahexaenoic acid ester of phloridzin inhibit lipopolysaccharide-induced inflammation in THP-1 differentiated macrophages. International Immunopharmacology. 25(1):199-206

[180] Mart´ın M A, Fern´andez-Mill´an E, Ramos s, Bravo L, and Goya L. (2014) Cocoa flavonoid epicatechin protects pancreatic beta cell viability and function against oxidative stress Molecular Nutrition & Food Research. 58(3):447-456

[181] Banihani S, Swedan S and Alguraan Z. (2013) Pomegranate and type 2 diabetes. Nutrition Research. 33(5):341-348

[182] Seymour E M, Lewis S K and Urcuyo-Llanes D E. (2009) Regular tart cherry intake alters abdominal adiposity, adipose gene transcription, and inflammation in obesity-prone rats fed a high fat diet. J Medicinal Food. 12(5):935-942

[183] Garcia-Diaz D F, Johnson M H and de Mejia E G. (2015) Anthocyanins from fermented berry beverages inhibit inflammation related adiposity response in vitro. J Medicinal Food. 18(4):489-496

[184] de la Garza A D, Etxeberria U and Palacios-Ortega S. (2014) Modulation of hyperglycemia and TNF-a-mediated inflammation by helichrysum and grapefruit extracts in diabetic db/db mice Food & Function. 5(9):2120-2128

[185] Y.-H. Chen, C.-H. Lee, Hsu T-H, and Lo H-C. (2015) Submerged culture mycelia and broth of the maitake medicinal mushroom *Grifola frondosa* (higher basidiomycetes) alleviate type 2 diabetes-induced alterations in immunocytic function. International J Medicinal Mushrooms. 17(6):541-556

[186] Li C, Ding Q, Nie S-P, Zhang Y-S, Xiong T, and Xie M-Y. (2014) Carrot juice fermented with *Lactobacillus plantarum* NCU116 ameliorates type 2 diabetes in rats. J Agricultural and Food Chemistry. 62(49):11884-11891

[187] Zhang Z, Wang S and Zhou S. (2014) Sulforaphane prevents the

development of cardiomyopathy in type 2 diabetic mice probably by reversing oxidative stress-induced inhibition of LKB1/AMPK pathway. J Molecular and Cellular Cardiology. 77:42-52

[188] Jung J Y, Lim Y, Moon M S, Kim J Y, and Kwon O. (2011) Onion peel extracts ameliorate hyperglycemia and insulin resistance in high fat diet/strepto zotocin-induced diabetic rats. Nutrition & Metabolism. 8(1):1-8

[189] Shin S, Lee S and Kwon J. (2009) Cordycepin suppresses expression of diabetes regulating genes by inhibition of lipopolysaccharide-induced inflammation in macrophages. Immune Network. 9(3):98-105

[190] Zhang F, Yang F, Zhao H, and An Y. (2015) Curcumin alleviates lung injury in diabetic rats by inhibiting nuclear factor-kB pathway. Clinical and Experimental Pharmacology and Physiology. 42(9):956-963

[191] Jim´enez-Flores L M, L´opez-Briones S, Mac´ıas-Cervantes M H, Ram´ırez-Emiliano J, and P´erez-V´azquez V. (2014) A PPAR gamma, NF-kB and AMPK-dependent mechanism may be involved in the beneficial effects of curcumin in the diabetic db/db mice liver. Molecules. 19(6):8289-8302

[192] Rastogi M, Ojha R P, Rajamanickam G V, Agrawal A, Aggarwal A, and Dubey G P. (2008) Curcuminoidsmodulates oxidative damage and mitochondrial dysfunction in diabetic rat brain. Free Radical Research. 42(11-12):999-1005

[193] Lin J, Zheng S and Chen A. (2009) Curcumin attenuates the effects of insulin on stimulating hepatic stellate cell activation by interrupting insulin signaling and attenuating oxidative stress. Laboratory Investigation. 89(12):1397-1409

[194] Pan Y, Zhu G and Y. W. (2013) Attenuation of high-glucose-induced inflammatory response by a novel curcumin derivative B06 contributes to its protection from diabetic pathogenic changes in rat kidney and heart,. J Nutritional Biochemistry. 24(1):146-155

[195] Kota N, Panpatil V V, Kaleb R, Varanasi B, and Polasa K. (2012) Dose-dependent effect in the inhibition of oxidative stress and anticlastogenic potential of ginger in STZ induced diabetic rats. Food Chemistry. 135(4):2954-2959

[196] Madkor H R, Mansour S W and Ramadan G. (2011) Modulatory effects of garlic, ginger, turmeric and their mixture on hyperglycaemia, dyslipidaemia and oxidative stress in streptozotocin-nicotinamide diabetic rats. British J Nutrition. 105(8):1210-1217

[197] Samarghandian S, Azimi-Nezhad M and Samini F. (2014) Ameliorative effect of saffron aqueous extract on hyperglycemia, hyperlipidemia, and oxidative stress on diabetic encephalopathy in streptozotocin induced experimental diabetes mellitus. Bio Med Research International. 2014:12 pages

[198] Altinoz E, Oner Z, Elbe H, Cigremis Y, and Turkoz Y. (2015) Protective effects of saffron (its active constituent, crocin) on nephropathy in streptozotocin-induced diabetic rats Human & Experimental Toxicology. 34(2):127-134

[199] Hazman O and Ovalı S. (2015) Investigation of the antiinflammatory effects of safranal on high-fat diet and multiple low-dose streptozotocin induced type 2 diabetes rat model. Inflammation. 38(3):1012-1019

[200] Campos K E, Diniz Y S, Cataneo A C, Faine L A, Alves Q F, and Novelli E L B. (2003) Hypoglycaemic and

antioxidant effects of onion, Allium cepa: dietary onion addition, antioxidant activity and hypoglycaemic effects on diabetic rats. International J Food Sciences and Nutrition. 54(3):241-246

[201] Bang M, Kim H.-A. and Cho Y-J. (2009) Alterations in the blood glucose, serum lipids and renal oxidative stress in diabetic rats by supplementation of onion (Allium cepa. Linn). Nutrition Research and Practice. 3(3):242-246

[202] Yokozawa T, Kim H Y, Cho E J, Yamabe N, and Choi J S. (2003) Protective effects of mustard leaf (Brassica juncea) against diabetic oxidative stress J Nutritional Science and Vitaminology. 49(2):87-93

[203] Mavondo G A and Musabayane C T. (2017) Transdermal Drug Delivery of Asiatic Acid Influences Renal Function and Electrolyte Handling in *Plasmodium berghei* - Infected Sprague-Dawley Male Rats. J Diseases and Medicinal Plants. 4(1):18-29.10.11648/j.jdmp.20180401.13

[204] Mavondo G A, Mkhwananzi B M, Mabandla M V, and Musabayane C T. (2016) ASIATIC ACID INFLUENCES GLUCOSE HOMEOSTASIS IN P. BERGHEI MURINE MALARIA INFECTED SPRAGUE-DAWLEY RATS. Afr J Tradit Complement Altern Med. 13(5):91-101.10.21010/ajtcam.v13i5.1391

[205] Boissier M-C, Semerano L, Challal S, Saidenberg-Kermanac'h N, and Falgarone G. (2012) Rheumatoid arthritis: fromautoimmunity to synovitis and joint destruction. Journal of Autoimmunity. 39(3):222-228

[206] Hitchon C A and El-Gabalawy H S. (2004) Oxidation in rheumatoid arthritis. Arthritis Research & Therapy. 6(6):265-278

[207] Fay j, Varoga D, Wruck c J, Kurz B, Goldring M B, and Pufe T. (2006) Reactive oxygen species induce expression of vascular endothelial growth factor in chondrocytes and human articular cartilage explants. Arthritis Research & Therapy. 8(article R189)

[208] McInnes I B and Schett G. (2007) Cytokines in the pathogenesis of rheumatoid arthritis. Nature Reviews Immunology. 7(6):429-442

[209] Yoo S-A, Kwok S-K and Kim W-U. (2008) Proinflammatory role of vascular endothelial growth factor in the pathogenesis of rheumatoid arthritis: prospects for therapeutic intervention Mediators of Inflammation. 2008(Article ID129873):6 pages

[210] Huber L C, Distler O, Tarner I, Gay R E, Gay S, and Pap T. (2006) Synovial fibroblasts: key players in rheumatoid arthritis. Rheumatology. 45(6):669-675

[211] Meinecke I, Rutkauskaite E, Gay S, and Pap T. (2005) The role of synovial fibroblasts in mediating joint destruction in rheumatoid arthritis. Current Pharmaceutical Design. 11(5):563-568

[212] Wruck C J, A. Fragoulis A and A. A G. (2011) Role of oxidative stress in rheumatoid arthritis: insights from the Nrf2-knockout mice. Annals of the Rheumatic Diseases. 70(5):844-850

[213] Wruck C J, Gotz M E, Herdegen T, Varoga D, Brandenburg L-O, and Pufe T. (2008) Kavalactones protect neural cells against amyloid B peptide-induced neurotoxicity via extracellular signal-regulated kinase 1/2-dependent nuclear factor erythroid 2-related factor 2 activation. Molecular Pharmacology. 73(6):1785-1795

[214] Jaiswal A K. (2004) Nrf2 signaling in coordinated activation of antioxidant gene expression. Free Radical Biology & Medicine. 36 (10):1199-1207

[215] Bu S Y, Lerner M and B.J. S. (2008) Dried plum polyphenols inhibit osteoclastogenesis by downregulating NFATc1 and inflammatory mediators. Calcified Tissue International. 82(6):475-488

[216] He Y-H, Zhou J and Wang Y-S. (2006) Anti-inflammatory and anti–oxidative effects of cherries on Freund's adjuvant-induced arthritis in rats. Scandinavian Journal of Rheumatology. 35(5):356-358

[217] Wahba M G F, Messiha B A S and Abo-Saif A A. (2015) Protective effects of fenofibrate and resveratrol in an aggressive model of rheumatoid arthritis in rats Pharmaceutical Biology. 54(9):1705-1715

[218] Tsubaki M, Takeda T and Kino T. (2015) Mangiferin suppresses CIA by suppressing the expression of TNF-a, IL-6, IL-1B, and RANKL through inhibiting the activation of NF-kB and ERK1/2. American J Translational Research. 7(8):1371-1381

[219] Luczkiewicz P, Kokotkiewicz A, Dampc A, and Luczkiewicz M. (2014) Mangiferin: a promising therapeutic agent for rheumatoid arthritis treatment. Medical Hypotheses. 83(5):570-574

[220] Yoon H-Y, Lee E-G and H. L. (2013) Kaempferol inhibits IL-1B-induced proliferation of rheumatoid arthritis synovial fibroblasts and the production of COX-2, PGE2 and MMPs International J Molecular Medicine. 32(4):971-977

[221] Gerhauser C, Klimo K, Heiss. E., Neumann I, Gamal-Eldeen A, Knauft J, et al. (2003) Mechanism-based in vitro screening of potential cancer chemopreventive agents. Mutat Res. 523-524:163-172

[222] Rasoanaivo P, Wright C W, Willcox M L, and Gilbert B. (2011) Whole plant extracts versus single compounds for the treatment of malaria: synergy and positive interactions. Malaria Journal. 10(1):S4.10.1186/1475-2875-10-S1-S4

[223] Coussens L M and Z. W. (2002) Inflammation and cancer. Nature. 420(6917):860-867

Chapter 6

Lipid Peroxidation and the Redox Effects of Polyherbal

Kale Oluwafemi Ezekiel

Abstract

The use of more than one herb in a medicinal preparation also known as polyherbal has increased geometrically in recent times. Over a hundred thousand scientists have cited "herbal" to strengthen its ethnopharmacological relevance in literature. Polyherbal (PH) is effective potential therapeutic compound used globally to treat oxidative stress-induced injuries which give credence for their traditional applications. However, some issues related to safety and adverse reactions due to PH have raised important public health debates. Lipid peroxidation (LPO) assay is widely used to assess the toxic endpoint of PH. This paper discusses some important roles that PH plays during oxidation–reduction processes.

Keywords: Polyherbal, Lipid Peroxidation, Oxidative Stress, Antioxidant, Ethnopharmacology

1. Introduction

Lipid peroxidation (LPO) is one of the oldest risk factors for oxidative stress and its mechanistic processes in disease modulation were first observed during the oxidative deterioration of edible materials [1]. The independent works of scientists in the identification of membrane shedding formed the earliest breakthrough for the study of lipid peroxidation [1, 2]. The biochemistry of oxygenase moiety was a light to the further understanding of the peroxidation of lipids [2]. The central roles play by small lipid molecules, free radicals and cytochrome p450 have been well established [3, 4]. The small lipid molecules are transformed into low-density lipoprotein (LDL). LDL modulations have contributed immensely to the studies on cellular biology. Both enzymatic and non-enzymatic approaches to scavenging LPO have been reported [5]. LPO has its root in several diseases including neurodegeneration, cardiovascular, respiratory, and cancers [2]. Consequently, the chemopreventive and protective roles of several antioxidants with proven efficacies have been documented in the literature. Although the different compounds of antioxidants have been identified, researchers have shown that it is by coordinated efforts that they can quench oxidative damage [6]. Evidence abounds of the involvement of LPO in several diseases and the protective roles of medicinal plant antioxidants [7, 8]. Often time in traditional medicine practices, medicinal compounds that perform a similar function are combined to obtain synergy [9, 10]. Thus, potential antioxidants are capable of preventing, protecting, and eliminating any form of identifiable oxidants [11]. The medicinal effects of functional food have been known for a long time. This observation was the result of preservative, anti-oxidative and antimicrobial actions found in ginger, nutmeg, turmeric, etc. These help arrest, repair,

IntechOpen

and restore potential injuries [12]. The scavenging efforts by antioxidants provide auto-inhibition or sometimes intervention to the activity of LPO metabolites [13]. Interestingly, in the case of progressive disease, the presence of combined efforts of antioxidants can offer a cascade of reactions to stop the process involved. Despite the tremendous roles of natural product medicines, some issues related to safety and adverse reactions due to PH have raised important public health debates [14]. Lipid peroxidation assay is one of the simplest methods widely used to assess the toxic endpoint of PH. Thus, this paper discusses some important roles that PH plays during oxidation–reduction processes.

2. Polyherbal (PH)

The use of herbal medicinal products has become popular for several reasons since they have always been the last order of resort when conventional therapy failed [15, 16]. Majorly, Polyherbal (PH, also known as herbal or traditional medicines) are used as complementary and alternative medicines [17]. The direct and indirect effects of polyherbal (PH) involvement in modulating LPO product formation have attracted scientific debates in recent times [15, 18]. Among the many reasons for use are that they are natural supplements, relatively safe, less toxic, and cheaper [15, 19]. The largest conventional therapy is obtained from plant sources, and many of them are nurtured within the neighborhood [20]. As the usual practice in traditional medicine, any substance or medicinal agent of natural origin may be used alone or with other agents as a polyherbal formulation which often requires less expertise [14]. PH is considered to be supplements and so boycotted major laws [21]. Thus, in most cases, measures for dose regulation are lacking in PH preparations as prohibition is difficult to reach [22]. This makes regulating agencies find it difficult to gather and provide the necessary implementation in many countries of the world. Despite the effort of the World Health Organization (WHO) to issue a certification scheme for the regulation of PH, the problem of non-compliance has been on the high side [23]. To this end, important PH like the *Aloe vera*, ginger, *Curcuma longa*, *Moringa oleifera*, *gingko biloba*, kava, milk thistle among others have birth beneficial compounds. Yet, all of these have been documented and updated for known adverse herb reactions [14, 22, 24]. Some variations in formulations among other factors could further influence the final herbal agents and/or products to be marketed. Reports on the adulteration of PH with conventional agents are grossly available and have raised serious concerns [22, 25]. Thus, the need for new regulations for botanicals has long been overdue. Other possibilities of PH being contaminated with undisclosed molecules around have been suggested to influence long-term consumption [26]. In respect, PH combination is a multifactorial antioxidant with the sole aim of scavenging oxidative radicals [16]. Both natural and synthetic antioxidants compounds abound [21, 27]. LPO products also result in deoxyribonucleic acid, cell membrane, and tissue damages in the human body [1, 2]. The global demand for antioxidants is evident and has prompted scientific interests in searching for new and safe antioxidant substances of natural origin. The WHO report shows both relevant authorities of people living in developing and developed countries have approved the use of traditional medicine for their primary healthcare. Plants possess the innate ability to synthesize a wide variety of enzymatic and non-enzymatic antioxidants capable of attenuating ROS-induced oxidative damage [12]. In respect, using appropriate separation techniques, we can now study the presence of phytoconstituents contributing to the antioxidant property of a given plant mixture [11, 16]. These mixtures may serve as a potential source of exogenous antioxidants to combat the undesirable effects of oxidative stress. Currently, this forms the basis for the PH that we now witnessed.

3. Importance of PH as Ethnomedicines

There are some conditions of inadequate efficacy, side effects, and pharmacokinetic problems of conventional drugs used for disease modulation [14]. Recent studies focused on the pharmacology and feasibility of herbal compounds as a potential strategy for alternative therapy. Now the discovery of novel therapeutic agents with multi-targeted potential is desirable [13, 20, 28]. Protective properties of phytochemicals combat numerous diseases and their vast acceptance and demand in human beings encouraged scientists to assess their effective activities [9, 11]. Artemisinin for instance has been known for decades and forms one of the most commonly prescribed medicinal agents [25]. The same has now emerged into an antimalarial drug used as first-line medication in this regard in addition to its adjuvant anticancer potential [10, 13]. Also, ethnomedicine has risen to provide an alternative to biofilm infections to improve medical treatments by the use of combinatorial treatment of bacterial biofilms as re-potentiators of classical antibiotics [9]. Actions related to anti-growth, anti-biofilm, or anti-quorum-sensing activities, to control bacterial infections have been associated with the use of PH including *P. granatum* or propolis compounds against known bacteria agents [9]. Other important derivatives of ethnomedicines are the phytoestrogens [8, 16]. They are a large family of plant-derived molecules possessing various degrees of estrogen-like activity; they exhibit agonist or antagonist estrogenic properties depending on the tissue. Delay of skin to undergo degenerative changes as it ages is now possible using relevant phytoestrogens knowledge to modulate skin elasticity, reductions in the epidermal thickness and collagen content, elastic fiber degeneration, and increased wrinkling and dryness. In turn, this is a key to senescence control via estrogen modulation [16]. Relevant interventions to target a variety of human diseases ranging from metabolic to brain disorders are now available [3, 12]. Several herbs have been reported to target the main biochemical events that are implicated in mental disorders, mimicking, to some extent, the mechanisms of action of conventional antidepressants and mood stabilizers with a wide margin of tolerability [3, 12]. With both experimental and clinical evidence, they rescue alterations in neurotransmitter and neuroendocrine systems, stimulate neurogenesis and the synthesis of neurotrophic factors, and they counteract oxidative stress, mitochondrial dysfunction, and inflammation e.g. saffron, crocin, etc. play a very significant role as a nutraceutical for cognitive functions affected by body injuries [11]. More so, ethnomedicine encourages nanotechnology development in the preparation of a herbometallic nano-drug [27, 29]. Recent studies on the physicochemical analysis confirmed that specific plant-derived herbometallic nano-drug such as rasa manikya nanoparticle were rich in mineral constituents and showed therapeutic opportunities for combating drug-resistant microbial strains among others [27]. Furthermore, some eminent components of traditional medicinal agents have contributed to cardio-metabolic disease treatment for decades. Phytomedicines such as berberine, lemon balm among others has attracted much interest for their pharmacological actions in managing cardio-metabolic diseases [29]. Recent discoveries of basic, translational and clinical studies have identified many novel molecular targets for phytocompounds, and provided novel evidence supporting the promising therapeutic potentials [16, 23, 26]. Hepatoprotective and renoprotective effects are two major medical challenges worldwide and a wide variety of herbs have been studied for the management of their related diseases. Bioactive compounds including silymarin, quercetin, curcumin, ginseng, and rutin for instance have long been used in traditional medicine [7]. Both in combating diseases, exerts hypolipidaemic and antioxidant effects, which prevents the fatty acid accumulation in the cells that may result from metabolic imbalances, and which affects multiple processes and signaling pathways [4, 13, 27].

4. Risk factors for PH-induced LPO

There are several setbacks to draw from the recent criticisms of PH applications [14, 25, 30]. This is the reason for the suggestions on standardization developments [15, 24, 28, 31]. The latter may help gain insight into their mechanisms as well as potentials for toxicity.

4.1 Dose

Several PH compounds have been extensively used as a traditional medicine for various therapies [18]. Because of the multifaceted component of PH mixture, several of the published articles showed that manufacturers relied mostly on the documented efficacies of the constituent compounds. There is a lack of specifications for the dose selection of subtherapeutic, therapeutic, and supratherapeutic doses which are used in animal studies, and for animal-to-man dose extrapolations [29]. Till now, there is a lack of formula to relate PH constituents; hence, lack of proper dose extrapolation of a single compound when combined with another may result in the potential mechanistic toxicological effect of the constituent mixture [28, 32]. Several misconceptions about dietary supplements being safe to have increased the number of hospital admissions [14, 17]. Recent studies on the PH dietary supplement popularly known as Cellgevity® (CG) confirmed that this premiere antioxidant supplement formula could act as a pro-oxidant, a substance capable of distorting the antioxidant systems [28]. Studies on the effects of therapeutic and supra-therapeutic doses of CG on reproductive function and biochemical indices in animals demonstrated some detrimental effects. CG is one of the most widely used glutathione supplements that has been considered to be harmless, nevertheless, this general assumption should not be overlooked. It is marketed to salvage for the body glutathione and/or complements its production. D-Ribose and L-Cysteine are the active compounds in CG in addition to the presence of vitamin C, selenium, alpha-lipoic, broccoli seed extract, curcumin, resveratrol, grape seed extract, quercetin, milk thistle seed extract, cordyceps, black pepper, aloe leaf. Some convergent opinions have highlighted the antioxidant activity of spices and their impact on human health, in particular, to increase reduced glutathione (GSH) concentration [12]. Whereas the presence of GSH improves protein function in a perturbed environment, its roles in modulating hypoxic apoptosis or oxidative stress is of great concern. This has put the faith of current supplement antioxidants in doubt, such as protein supplements and others, which now beg for safety evaluation. Previous studies have reported that dietary supplements are now being used to prevent and treat various diseases [7]. We have now understood that any excess of antioxidants could be detrimental and can result in adverse events and even death. Reports over the past few years have implicated the use of herbs and herbal products to generate reactive oxygen and nitrogen species. Mahwangyounpae-tang (MGT) is another but very popular antioxidant PH consisting of about 22 compounds [32]. Reports on MGT showed that MGT extract was safe for use in asthma at the sub-acute repeated oral dose levels. However, when the doses were increased per daily in rodents, there were correspondingly increased morphological aberrations and organ histoarchitectural changes characterized by hypertrophy of the heart and tubular necrosis of the kidney [32]. Thus, concerns over the doses of antioxidants when using alone, or with other drugs have arisen (**Table 1**).

4.2 Duration of treatment

Evaluations of most of the PH for toxicological profiles are most unlikely not going to be sufficient to determine the endpoint toxicity [24]. For instance, the majority of the acute to chronic studies found in the literature were between 14 and 90 or 180 days. However, only a few studies show reversibility assessments or show species specifications [38]. Over 80–90% of the toxicological studies published showed few or no adverse drug reactions. Also, the parameters for their toxicological endpoint did not capture genotoxicity potential [39]. As with the case of CG, which supposedly should provide a maximum antioxidant function by GSH synthesis rates and concentrations as expressed by the content could act as a pro-oxidant causing oxidative damage in normal humans particularly at high dosages [28]. Such transition in chemical nature could generate pro-oxidant–antioxidant imbalance thereby produce undesirable toxins leading to oxidative stress [22, 25]. Hence, a long-term toxicological profile plus storage history has been recommended to ascertain PH safety.

Polyherbal	Uses	Potential Mechanisms	Potential Target
Mahwangyounpae-tang® [6]	anti-inflammatory and asthma	Immune system, blood, pro-inflammation	Lungs, heart and kidney
Cellgevity® [28]	Antioxidant supplement formula	Immune system, haematotoxicity, pro-inflammation, elevation of serum low density lipoprotein	Blood, liver, testis, kidneys
Hydroxycut [30]	Weight loss supplement	Elevation of hepatic biomarkers, pro-inflammation	Liver
Bon-sante cleanser® [33]	Body hormones booster and energizer	Immune system and body hormones, pro-inflammation	Liver, heart
Bronco-T® [34]	Anti-inflammatory and lung regeneration	Immune system, blood, pro-inflammation, bronchoconstriction	Lungs
Jambadyarista® [35]	Diabetes and its associated complications	Hypolipidic, Nuclear factor-κappa B activation	Pancreas
Shengmai formula® [36]	Cardiovascular diseases	Opening of mitochondrial permeability transition pore, cyclophilin D.	Heart, liver
Hab-e-Kabad Noshadri Hepatic disorders® [37]	Hepatic disorders, abdominal problems.	Elevation of hepatic and renal biomarkers, inflammation	Liver, kidneys and gastrointestinal system

Table 1.
Shows the potential targets of PH-induced lipid peroxidation.

4.3 Lethal dose estimation

The median lethal dose (LD_{50}) is the statistically derived dose following the administration of any PH which is expected to produce death in 50% of the treated population [40]. The toxic effects of chemicals, food substances, pharmaceuticals, etc., have attained great significance in the 21st century [17]. Toxicity tests are mostly used to examine specific adverse events or specific endpoints in disease identifications. Toxicity testing also helps to calculate the No or Low Observed Adverse Effect Level (NOAEL/LOAEL) dose and is helpful for clinical studies [39]. However, the methods of determination of median lethal dose (LD50) may impact negatively on the information for the use of PH. Therefore exaggerated lipid peroxidation levels due to PH might be the results of poor safety methodology [23]. This however can be minimized by using several methods to ascertain the PH LD50 level. Studies have suggested that lack of expertise in this aspect of scientific investigation might influence judgment on the PH formula [24]. Inability to in-cooperate the knowledge of individual median lethal dose might create impurities in the constituent mixture. The five most commonly paraded bitters in Nigeria for type 2 diabetes are Yoyo bitters (YB), Oroki herbal mixture (OB), Ruzu Bitters (RB), Fijk flusher (FB), and Fidson Bitter (FB) respectively. Although, each of this preparation has claimed for several indications, however, scientific investigations for the mixture have reported weight trimming and blood sugar modulations (Kale et al., 2018). Since their constituents are known, their ability to act synergistically and the potential for precipitating adverse herb reaction of LPO have been ascertained [15]. These concerned mixtures have multifaceted constituents most of which have popular applications. Examples include ginseng, *Aloe vera*, *Citrus aurantifolia*, *Sorghum bicolor*, *Mangifera indica*, etc. These have yielded very important phytomedicines or bioactive components that have been confirmed by different studies, although, some exhibit overlapping effects in disease management. *Chenopodium murale* (Chenopodiaceae) has yielded analgesics, anti-inflammatory, anti-fungal, anti-bacterial, anti-oxidant, hypotensive, and hepatoprotective molecules. These PH administrations improved lipid parameters in diabetic rats. While hypercholesterolemia persists in diabetic rats treated with RB and FJB, RB increased LDL in treated rats [15]. Further, RB, FJB, and OB showed tendencies to elevate serum TC while RB increases LDL cholesterol in rats. This indicates the suitability of these products to produce LPO molecules as a risk factor for dyslipidemia in potential users.

4.4 Manufacturer Bias

In contrast to the general belief, medicinal plant preparations have been shown to pose serious health risks in a dose and time-dependent either alone or in combination with other agents [19, 38]. Reports have it that PH contain tightly bound bioactive compounds and have shown the possibilities of an indirect risk that can be independent of the active compound [16]. Countries are now examining national pharmacovigilance data using statistical tools to report population possible risks [24]. This is because, oftentimes, the species specification by manufacturers on PH constituents did not correlate most times with products packaging thereby altering scientific decision on toxicological evaluations [25]. The potential to generate lipid peroxidation by several PH are underestimated [14]. On the other hand, the latter creates a contrast in mind when a single compound has a toxicity level greater than the combined mixture. Additionally, erroneous claims of total cure and or weight supporting supplement some of these PH are marketed with approved consent from regulatory authorities [22]. There are also exaggerations of PH with conventional drugs which have been suggested to contribute in part to the unwanted adverse reaction of some PH [14, 20, 23]. Thus, this issue associated with production bias has raised concern about quality control, screening methods as well as toxicity scoring which most regulatory authorities however have not been able to properly address it. Bon-santé cleanser® (BS) is a popular PH comprising of anogeissus leiocarpus (DC., family Combretaceae), *Terminalia ivorensis* (A. Chev.), massularia acuminate (G. Don,) Bullock ex Hoyle and macuna pruriens (L.,) DC (fabaceae) respectively which have been formulated into capsule. The proposed claims include androgenic, antipyretic, analgesic, and anti-inflammatory potentials. The pharmacological activities of *M. acuminata*, *T. ivorensis*, A. leiocarpus and *M. pruriens* have been adduced to be due to the presence of glycosides, dimeric antioxidants, phenolics and flavones respectively. Despite the relevance of BS in boosting the body hormones including the follicle-stimulating, luteinizing hormone and testosterone respectively, reports have implicated this BS as a potential hepatotoxicant and a commodity with pro-oxidants status. Increasing the dose of BS was associated with significantly reduced sperm motility, live-dead ratio, testis weight, and cause mild inflammation in the vital organs testis in the animals [40]. There was diminishing return as higher doses did not exert any significantly different change in the level of body hormones which could cause the negative feedback effect on the anterior pituitary [33]. Further, this PH demonstrates the potential to induce LPO of the testicular origin or promote the generation of free radicals in vivo.

5. Molecular mechanisms of PH-induced LPO

Acridocarpus smeathmannii (DC.) is widely used for the treatment of male infertility, anemia, and pains in traditional medicine as well as an herbal mixture. *A. smeathmannii* has gained popularity particularly worldwide for the management of infertility, anemia, pain, and some cutaneous infections [20, 41]. Studies have demonstrated the ethnobotanical relevance of this plant. Thus, recent reports demonstrated the bioactive compounds, sexual behavior, and associated reproductive function via biochemical and pharmacological mechanisms due to the hydroethanolic root extract of *A. smethmannii* (HAS) and subchronic oral toxicity effects of HAS [31]. Also, the organization for economic co-operation and development (Test No. 453) [39] has approved that very long-term toxicological and/or carcinogenicity studies be carried, in particular for PH that are considered not to be potentially toxic at therapeutic and supratherapeutic doses. In this study, the possible systemic toxicological changes following a 180 days administration in Wistar rats of both sexes under approved guidelines for animal use following the procedures as documented by Kilkenny et al. [42] for reporting animal research. Animals (Wistar rats, male: female = 1:1 = 48) received distilled water (10 mL/kg) or HAS (250, 500 or 1000 mg/kg body weight per day) consecutively for 180 days. From the results obtained, HAS (500 and 1000 mg/kg) demonstrates such a tendency to increase oxidative stress parameters via an increase in LPO as malondialdehyde (MDA) levels in vital organs (**Figure 1**) in rats. Although, evidence abounds of the turnover of products of LPO in the body, however, the presence of an adverse reaction may become aggravated time-dependently or produces an interaction with other substances presences [14, 26]. This could impose either a self or even exaggerates the effect of HAS in a given period. Increasing levels of the intracellular antioxidants, GSH levels, could not overcome the LPO products induced by the highest dose of extract in rats, thus, highlighting potential adverse effects of HAS in vivo. This suggests that PH can induce LPO metabolites such as MDA which by this observation could translate into a clinically relevant situation. More so, oestrogen level was reduced in treated female rats as obtained in these results. Also, in both sexes, HAS (500 mg/kg) and HAS (1000 mg/kg) showed elevated serum nitric oxide (NO) levels respectively (**Table 2**).

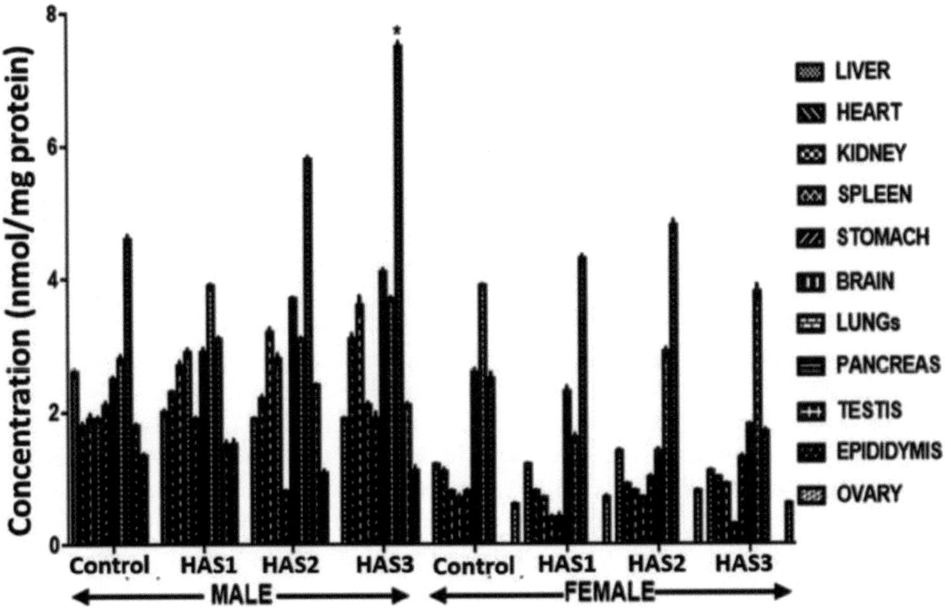

Figure 1.
Effect of HAS on lipid peroxidation in normal Wistar rats. Results are expressed as mean ± S.E.M. n: Total number per group. n = 12. Mortality = HAS (500) (male, 25%), HAS (1000) (male, 33.3%), HAS1 (250 mg/kg) (female, 8.3%), HAS2 (500 mg/kg) (female, 25%) and HAS (1000 mg/kg) (female, 41.7%) respectively. $^{*}p < 0.05$ *or* $^{**}p < 0.01$ *compared with control (distilled water: DW, 10 mL/kg) group. HAS1 (250): 250 mg/kg, HAS2 (500): 500 mg/kg, HAS3 (1000): 1000 mg/kg, HAS: Hydroethanolic extract of* Acridocarpus smeathmannii *root.*

	Control	HAS1 (250 mg/kg)	HAS2 (500 mg/kg)	HAS3 (1000 mg/kg)
Testosterone^{x}	3.56 ± 0.09	4.53 ± 0.04*	4.84 ± 0.06*	6.48 ± 0.13*
PSA^{x}	1.83 ± 0.01	1.89 ± 0.01	1.81 ± 0.01	0.69 ± 0.02*
Nitric Oxide^{x}	0.65 ± 0.10	0.85 ± 0.10	1.28 ± 0.10*	1.83 ± 0.20*
TNF-α^{x}	3.19 ± 0.19	2.72 ± 0.19	4.09 ± 0.18	4.46 ± 0.20*
NF-kB^{x}	1.03 ± 0.14	1.64 ± 0.16	2.02 ± 0.13*	2.83 ± 0.12*
Oestrogen^{y}	51.72 ± 1.43	63.81 ± 2.26	51.44 ± 1.38	40.55 ± 2.13*
Progesterone^{y}	43.91 ± 0.39	125.03 ± 0.58*	143.91 ± 0.29**	168.60 ± 0.38**
Nitric Oxide^{y}	1.40 ± 0.01	1.38 ± 0.01	2.50 ± 0.02*	2.40 ± 0.02*
TNF-α^{y}	5.12 ± 0.19	6.90 ± 0.18	6.79 ± 0.19	8.44 ± 0.18*
NF-kB^{y}	1.25 ± 0.11	1.32 ± 0.14	3.23 ± 0.12*	4.11 ± 0.16*

Results are expressed as mean ± S.E.M. n = 12. HAS: hydroethanolic extract of Acridocarpus smeathmannii root. $^{*}p < 0.05$ *or* $^{**}p < 0.01$ *compared with control distilled water group. "x" and "y" in superscript represented "male" and "female" rats respectively. Mortality: HAS (500) (male, 25%), HAS (1000) (male, 33.3%), HAS (250) (female, 8.3%), HAS (500) (female, 25%) and HAS (1000) (female, 41.7%). Oestrogen (pg/mL), Progesterone (ng/mL), Testosterone (ng/mL), Serum Nitric Oxide (nmol/mL), TNF-α: Tumor Necrosis Factor-alpha, PSA: Prostate Specific Antigen (ng/mL). Differences between groups were determined by one-way analysis of variance (ANOVA) using Statistical Package for Social Sciences (SPSS, version 20.0) software for windows and Post hoc test for intergroup using the least significant difference, followed by Dunnett's test. Significance was considered at p < 0.05. All results were expressed as the mean ± standard error of the mean.*

Table 2.
Effect of hydroethanolic root of Acridocarpus smeathmannii *extract on body hormones and molecular biomarkers in serum of normal and treated rats using enzyme-linked immunosorbent assays.*

Nitric oxide plays a very significant role as a signaling molecule in several biological events [43]. Also, the highest dose, HAS (1000 mg/kg) increased serum tumor necrosis factor-alpha (TNF-α) level, a hallmark biomarker of tissue injury, in rats of both sexes (**Table 2**). The presence of TNF-α level could indicate some processes involving the action of nuclear factor kappa-B and a host of other mediators of injury [44]. Even at the lowest dose (250 mg/kg) of HAS administered to rats, serum NF-kB levels increase (**Table 2**). The aforementioned provides such suitability that HAS administration may induce cellular changes ranging from inflammatory reactions to LPO metabolites. An inducible nitric oxide synthase (iNOS) is an importantly upregulated isoform of NOS that synthesis the inducible nitric oxide (iNO) [45]. iNO is a disease-causing agent that can increase iNOS mRNA, protein, and then causes activity that may significantly alter NO turnover [42]. From the study results, administration of HAS after 180 days increases the chances of iNOS expression (**Figure 2**) in rats thereby showing the potential inherent pro-inflammatory and pro-oxidant properties in HAS that can cause tissue damage.

5.1 MDA levels

HAS1 lowered MDA level in the stomach by 32.2% in normal male rats (**Figure 1**). However, HAS2 produces an increased ($p < 0.05$) MDA levels in pancreas, heart, and

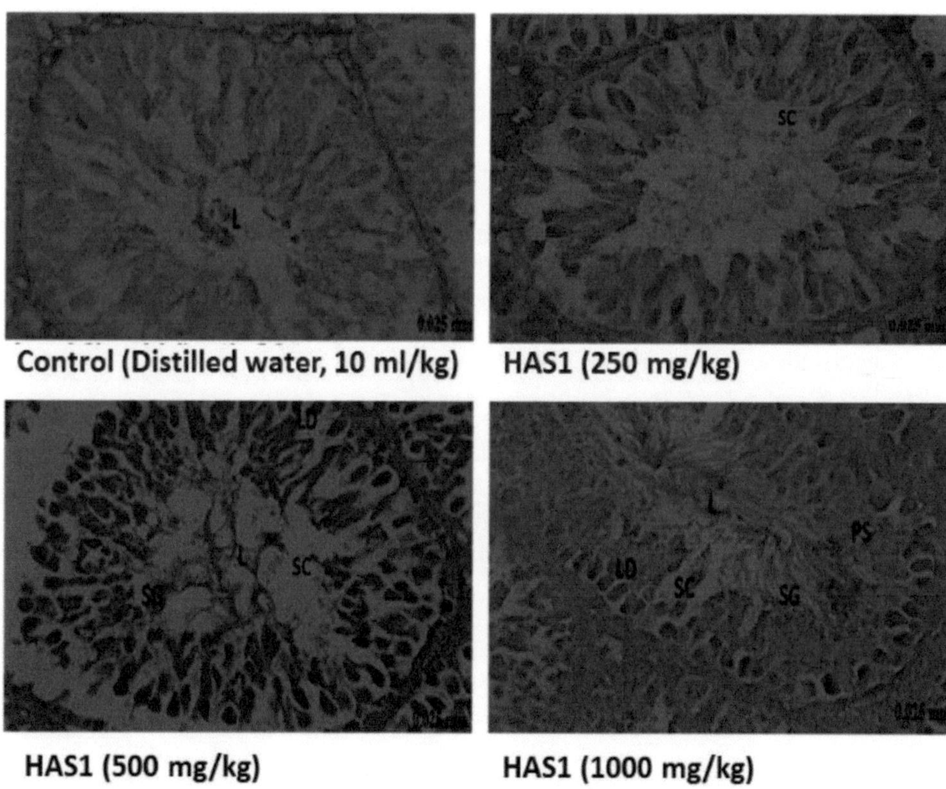

Figure 2.
Immunohistochemical photomicrograph of inducible nitric oxide synthase (iNOS) staining in the testis of male rats. The whole germ layer in control (Normal saline, 10 mL/kg) group show iNOS negative immunoreactivity. The Sertoli cells (arrows) majorly show positive reaction in HAS1 (250 mg/kg) group while the whole germ cells in the HAS2 (500 mg/kg) group have positive reactions. The primary spermatogonia show positive reaction than other cells in the germinal layer in the HAS3 (1000 mg/kg) (arrows). SG, Spermatogonium; SC, Sertoli cell; PS, primary spermatocyte; L, lumen. HAS, Hydroethanolic extract of Acridpcarpus smathmannii *root. (Mag. ×400.)*

brain by 113.64%, 70.37%, and 36% respectively. Additionally, HAS3 elevated MDA levels in testis (79.41%), kidney (52.38%), pancreas (104.55%), heart (88.89%), and brain (96%) respectively in rats. on the other hand, in the female rats, HAS1 caused an increase in MDA level in lung (57.69%) and pancreas (88%) respectively. Similarly, HAS2 produce elevated MDA levels in the lung and pancreas by 80.77% and 44% in the treated rats, whereas, HAS3 further increased ($p < 0.05$) MDA levels in the lung (50%), brain (82.14%), and pancreas (24%) respectively in the treated animals.

5.2 Reproductive hormones and molecular biomarkers

HAS3, HAS2 and HAS1 showed a dose-dependently decrease ($p < 0.05$) testosterone levels by 82.02%, 36.01% and 27.25% respectively (**Table 2**). Additionally, the PSA level was lowered in rats that received HAS3 by 62.44%. On the other hand, in female rats, HAS1 and HAS2 administration elevated serum oestrogen levels by 23.38%, whereas HAS2 and HAS3 lowered oestrogen levels by 0.54% and 21.60% when compared with control. Further, progesterone was increased in all the treated female rats by 283.97% (HAS3), 227.74% (HAS2), and 184.74% (HAS1) respectively. In male rats, serum NO levels were elevated ($p < 0.05$) following HAS2 and HAS3 administrations. Also, HAS3 increased ($p < 0.05$) TNF-α level by 39.99% in rats. In the female rats, however, both HAS2 and HAS3 elevated ($p < 0.05$) serum NO levels by 78.57% and 71.43% respectively. Further, TNF-α and NF-kB were increased in rats serum of HAS3 by 64.84% and 228.8% when compared with control.

6. Conclusions

The evidence that complementary and alternative therapy plays a crucial role in the management of health is indisputable. However, because of the complexity of the phytocompound present in PH, they may act as pro-oxidants as well as antioxidants. Therefore, as pro-oxidants, they generate lipid peroxidation products which are an important risk factors for tissue damage.

Acknowledgements

The kind leadership of my teachers and mentors, Dr. Awodele O. and Dr. Akindele A. J., of Department of Pharmacology, Therapeutics and Toxicology, College of Medicine, University of Lagos, Lagos Nigeria are gratefully acknowledged.

Conflict of interest

Authors declare that they have no competing interests. No specific grant received from any agency in the public, commercial, or not-for-profit sectors.

Acronyms and abbreviations

LPO	Lipid Peroxidation
MDA	Malondialdehyde

PH	Polyherbal
TNF-α	Tumor Necrosis Factor-alpha
HAS	Hydroethanolic extract of *Acridocarpus smeathmannii* root
iNOS	Inducible nitric oxide synthase
LD_{50}	Median lethal dose
ROS/RNS	Reactive Oxygen/Nitrogen Species

Author details

Kale Oluwafemi Ezekiel[1,2]

1 Department of Pharmacology, School of Medicine, Babcock University, Ilishan-Remo, Ikeja, Ogun State, Nigeria

2 Department of Pharmacology, Therapeutics and Toxicology, Obafemi Awolowo College of Health Sciences, Sagamu Campus, Olabisi Onabanjo University, Ogun State, Nigeria

*Address all correspondence to: kalefemi@gmail.com;
kale.oluwafemi@oouagoiwoye.oou.edu.ng

References

[1] Macho-González, A., Garcimartín, A., López-Oliva, M. E., Bastida, S., Benedí, J., Ros, G., ... & Sánchez-Muniz, F. J. (2020). Can Meat and Meat-Products Induce Oxidative Stress? Antioxidants, *9*(7), 638

[2] Girotti, A. W. (2001). Photosensitized oxidation of membrane lipids: reaction pathways, cytotoxic effects, and cytoprotective mechanisms. Journal of Photochemistry and Photobiology B: Biology, *63*(1-3), 103-113

[3] Halliwell, B. (2001). Role of free radicals in the neurodegenerative diseases. Drugs & aging, *18*(9), 685-716.

[4] Panigrahy, D., Kaipainen, A., Greene, E. R., & Huang, S. (2010). Cytochrome P450-derived eicosanoids: the neglected pathway in cancer. Cancer and Metastasis Reviews, *29*(4), 723-735.

[5] Niki, E., Yoshida, Y., Saito, Y., & Noguchi, N. (2005). Lipid peroxidation: mechanisms, inhibition, and biological effects. Biochemical and biophysical research communications, *338*(1), 668-676

[6] Park, M.Y., Choi, H.Y., Kim, J.D., Lee, H.S., Ku, S.K. 28 Days repeated oral dose toxicity test of aqueous extracts of mahwangyounpae-tang, a polyherbal formula. Food Chem Toxicol. 2010 Aug-Sep;48(8-9):2477-2482. doi: 10.1016/j.fct.2010.06.017. Epub 2010 Jun 15. PMID: 20558228

[7] Volate, S. R., Davenport, D. M., Muga, S. J., & Wargovich, M. J. (2005). Modulation of aberrant crypt foci and apoptosis by dietary herbal supplements (quercetin, curcumin, silymarin, ginseng and rutin). Carcinogenesis, *26*(8), 1450-1456.

[8] Liu, T., Li, N., Yan, Y. Q., Liu, Y., Xiong, K., Liu, Y., ... & Liu, Z. D. (2020). Recent advances in the anti-aging effects of phytoestrogens on collagen, water content, and oxidative stress. Phytotherapy Research, *34*(3), 435-447.

[9] Atanasov, A.G., Waltenberger, B., Pferschy-Wenzig, E.M., Linder, T., Wawrosch, C., Uhrin P., Temml, V., Wang, L., Schwaiger, S., Heiss, E.H., Rollinger, J.M., Schuster, D., Breuss, J.M., Bochkov, V., Mihovilovic, M.D., Kopp, B., Bauer, R., Dirsch, V.M., Stuppner, H. Discovery and resupply of pharmacologically active plant-derived natural products: A review. Biotechnol Adv. 2015 Dec;33(8):1582-1614. doi: 10.1016/j.biotechadv.2015.08.001. Epub 2015 Aug 15. PMID: 26281720; PMCID: PMC4748402.

[10] Jaiswal, Y., Liang, Z., & Zhao, Z. (2016). Botanical drugs in Ayurveda and traditional Chinese medicine. Journal of ethnopharmacology, *194*, 245-259

[11] Howes, M. J. R., Perry, N. S., Vásquez-Londoño, C., & Perry, E. K. (2020). Role of phytochemicals as nutraceuticals for cognitive functions affected in ageing. British journal of pharmacology, *177*(6), 1294-1315.

[12] Yashin, A., Yashin, Y., Xia, X., & Nemzer, B. (2017). Antioxidant activity of spices and their impact on human health: A review. Antioxidants, *6*(3), 70.

[13] Goyal, S., Gupta, N., Chatterjee, S., & Nimesh, S. (2017). Natural plant extracts as potential therapeutic agents for the treatment of cancer. Current topics in medicinal chemistry, *17*(2), 96-106.

[14] Ekor, M. (2014). The growing use of herbal medicines: issues relating to adverse reactions and challenges in monitoring safety. Frontiers in pharmacology, *4*, 177.

[15] Kale, O. E., Akinpelu, O. B., Bakare, A. A., Yusuf, F. O., Gomba, R., Araka, D. C., ... & Odutola, O. (2018). Five traditional Nigerian Polyherbal remedies protect against high fructose fed, Streptozotocin-induced type 2 diabetes in male Wistar rats. BMC complementary and alternative medicine, *18*(1), 1-11

[16] Widjajakusuma, E. C., Jonosewojo, A., Hendriati, L., Wijaya, S., Surjadhana, A., Sastrowardoyo, W., ... & Esar, S. Y. (2019). Phytochemical screening and preliminary clinical trials of the aqueous extract mixture of Andrographis paniculata (Burm. f.) Wall. ex Nees and Syzygium polyanthum (Wight.) Walp leaves in metformin treated patients with type 2 diabetes. Phytomedicine, *55*, 137-147.

[17] Clarke, T. C., Black, L. I., Stussman, B. J., Barnes, P. M., & Nahin, R. L. (2015). Trends in the use of complementary health approaches among adults: United States, 2002-2012. National health statistics reports, (79), 1.

[18] Hur, S. J., Kang, S. H., Jung, H. S., Kim, S. C., Jeon, H. S., Kim, I. H., & Lee, J. D. (2012). Review of natural products actions on cytokines in inflammatory bowel disease. Nutrition research, *32*(11), 801-816

[19] Awodele, O., Popoola, T. D., Amadi, K. C., Coker, H. A. B., & Akintonwa, A. (2013). Traditional medicinal plants in Nigeria—Remedies or risks. Journal of ethnopharmacology, *150*(2), 614-618.

[20] Catarino, L., Havik, P. J., & Romeiras, M. M. (2016). Medicinal plants of Guinea-Bissau: Therapeutic applications, ethnic diversity and knowledge transfer. Journal of ethnopharmacology, *183*, 71-94.

[21] Cordell, G. A., & Colvard, M. D. (2012). Natural products and traditional medicine: turning on a paradigm. Journal of natural products, *75*(3), 514-525.

[22] Binns, C. W., Lee, M. K., & Lee, A. H. (2018). Problems and prospects: public health regulation of dietary supplements. Annual review of public health, *39*, 403-420.

[23] Enioutina, E. Y., Salis, E. R., Job, K. M., Gubarev, M. I., Krepkova, L. V., & Sherwin, C. M. (2017). Herbal Medicines: challenges in the modern world. Part 5. status and current directions of complementary and alternative herbal medicine worldwide. Expert review of clinical pharmacology, *10*(3), 327-338.

[24] Gromek, K., Drumond, N., & Simas, P. (2015). Pharmacovigilance of herbal medicines. International Journal of Risk & Safety in Medicine, *27*(2), 55-65.

[25] Ernst, E. (2002). Adulteration of Chinese herbal medicines with synthetic drugs: a systematic review. Journal of internal medicine, *252*(2), 107-113.

[26] Mouly, S., Lloret-Linares, C., Sellier, P. O., Sene, D., & Bergmann, J. F. (2017). Is the clinical relevance of drug-food and drug-herb interactions limited to grapefruit juice and Saint-John's Wort?. Pharmacological research, *118*, 82-92.

[27] Ruidas B, Som Chaudhury S, Pal K, Sarkar PK, Das Mukhopadhury S. A novel herbometallic nanodrug has a potential for antibacterial and anticancer through oxidative damage. Nanomedicinen(Lond). 2019; 14(9):1173-1189.

[28] Awodele O, Badru WA, Busari AA, Kale OE, Ajayi TB, Udeh RO, Emeka PM. Toxicological evaluation of therapeutic and supra-therapeutic doses of Cellgevity® on reproductive function and biochemical indices in Wistar rats. BMC Pharmacol Toxicol. 2018 Oct

25;19(1):68. doi: 10.1186/s40360-018-0253-y).

[29] Belwal, T., Ezzat, S. M., Rastrelli, L., Bhatt, I. D., Daglia, M., Baldi, A., ... & Atanasov, A. G. (2018). A critical analysis of extraction techniques used for botanicals: Trends, priorities, industrial uses and optimization strategies. TrAC Trends in Analytical Chemistry, *100*, 82-102.

[30] Dara, L., Hewett, J., & Lim, J. K. (2008). Hydroxycut hepatotoxicity: a case series and review of liver toxicity from herbal weight loss supplements. World journal of gastroenterology: WJG, *14*(45), 6999.

[31] Kale OE, Awodele O, Akindele AJ. Acridocarpus Smeathmannii (DC.) Guill. & Perr. Root enhanced reproductive behavior and sexual function in male wistar rats: Biochemical and pharmacological mechanisms. J Ethnopharmacol. 2019 Feb 10;230:95-108. doi: 10.1016/j.jep.2018.10.024. Epub 2018 Oct 31. PMID: 30389468.

[32] Park MY, Choi HY, Kim JD, Lee HS, Ku SK. 28 Days repeated oral dose toxicity test of aqueous extracts of mahwangyounpae-tang, a polyherbal formula. Food Chem Toxicol. 2010 Aug-Sep;48(8-9):2477-2482. doi: 10.1016/j.fct.2010.06.017. Epub 2010 Jun 15. PMID: 20558228.

[33] Awodele O., Kale OE, Odewabi AO, Ekor M, Salau BA, Adefule-Ositelu AO. Safety evaluation of Bon-santé cleanser® polyherbal in male Wistar rats: Further investigations on androgenic and toxicological profile. J Tradit Complement Med. 2017 Jun 20;8(1):212-219. doi: 10.1016/j.jtcme.2017.06.002. PMID: 29322011; PMCID: PMC5756022)

[34] Sholapuri, P., Chintha, V., Matcha, B., & Pradeepkiran, J. (2020). Beneficial effects of polyherbal formulation (Bronco-T) on formaldehyde-induced lung toxicity in male Wistar rats. Toxicology Research, *9*(6), 798-807.

[35] Hasan M, Mahmud AA, Alam MJ, Siddiqui SA, Arman MSI, Mahmud MH, Amin MN, Imtiaz O, Shahriar M, Jakaria M. Subacute oral toxicity of ayurvedic anti-diabetic preparation *Jambadyarista* in Sprague-Dawley rats. Toxicol Rep. 2020 Dec 1;7:1616-1621. doi: 10.1016/j.toxrep.2020.11.011. PMID: 33318950; PMCID: PMC7725955.

[36] Li L, Yang D, Li J, Niu L, Chen Y, Zhao X, Oduro PK, Wei C, Xu Z, Wang Q, Li Y. Investigation of cardiovascular protective effect of Shenmai injection by network pharmacology and pharmacological evaluation. BMC Complement Med Ther. 2020 Apr 15;20(1):112. doi: 10.1186/s12906-020-02905-8. PMID: 32293408; PMCID: PMC7158159.

[37] Ishtiaq S, Akram M, Kamran SH, Hanif U, Afridi MSK, Sajid-Ur-Rehman, Afzal A, Asif A, Younus M, Akbar S. Acute and sub-acute toxicity study of a Pakistani polyherbal formulation. BMC Complement Altern Med. 2017 Aug 4;17(1):387. doi: 10.1186/s12906-017-1889-7. PMID: 28778156; PMCID: PMC5545041.

[38] Akindele, A. J., Unachukwu, E. G., & Osiagwu, D. D. (2015). 90 Days toxicological assessment of hydroethanolic leaf extract of Ipomoea asarifolia (Desr.) Roem. and Schult. (Convolvulaceae) in rats. Journal of ethnopharmacology, *174*, 582-594.

[39] Organisation for Economic Co-operation and Development. (2009). *Test No. 453: Combined Chronic Toxicity/Carcinogenicity Studies*. OECD Publishing.

[40] Kale OE, Awodele O. Safety evaluation of Bon-santé cleanser® polyherbal in male Wistar rats. BMC Complement Altern Med. 2016 Jul

7;16:188. doi: 10.1186/
s12906-016-1188-8.

[41] Van Andel, T. R., Croft, S., Van Loon, E. E., Quiroz, D., Towns, A. M., & Raes, N. (2015). Prioritizing West African medicinal plants for conservation and sustainable extraction studies based on market surveys and species distribution models. Biological Conservation, *181*, 173-181.

[42] Kilkenny C, Browne W, Cuthill IC, Emerson M, Altman DG. Animal research: reporting in vivo experiments—The ARRIVE Guidelines. Journal of Cerebral Blood Flow & Metabolism. 2011; 31(4):991.

[43] Ignarro, L. J. (2019). Nitric oxide is not just blowing in the wind. Br J Pharmacol., 176(2): 131 – 134.

[44] Skotheim, R. I., Monni, O., Mousses, S., Fosså, S. D., Kallioniemi, O. P., Lothe, R. A., & Kallioniemi, A. (2002). New insights into testicular germ cell tumorigenesis from gene expression profiling. Cancer research, *62*(8), 2359-2364.

[45] Aktan, F. (2004). iNOS-mediated nitric oxide production and its regulation. Life sciences, *75*(6), 639-653.

Chapter 7

Lipid Peroxidation: Aging Kidney

Harnavi Harun

Abstract

Kidney is one of the tissues affected by age that involves cellular and structural changes inside the kidney and notably implicates with comorbidity, related to cardiovascular disease aging. Aging kidney causes the elderly susceptible to clinical deterioration from ordinary stimulation that younger individual can compensate, including acute renal injury, volume depletion or overload, sodium and potassium level disorders, and toxic reaction against kidney excreted drugs. As one of the organs with the fastest aging rate, kidney shows several age-related decline in both structural and functional with 30% of the glomerulus are damaged and represent diffuse glomerular sclerosis by age 75 and explain why the prevalence of chronic kidney disease (CKD) and end-stage renal disease are very common in the elderly. The cross-sectional population-based study by The National Health and Nutrition Examination Survey supports the theory of age-related decline in kidney function, although some other subjects did not have an absolute decline in kidney function. The underlying molecular mechanisms could be the target of future therapeutic strategies. Aging is a natural biological process characterized by a gradual decline in cellular function as well as progressive structural change of organ systems. In aging kidney, there are interactions of genetic factors, environmental changes, and cellular dysfunction that lead to the typical structural and functional changes. One of the most popular theory of aging is the theory of free radicals or oxidative stress based on the fact that cells are under chronic oxidative stress due to an imbalance between pro oxidants and antioxidants. Reactive oxygen species are oxygen-derived oxidizing compounds that are highly reactive, consisting of free radicals and non-radicals. Reactive oxygen species (ROS) and reactive nitrogen species (RNS) refer to both reactive radicals and non-radical derivatives of oxygen and nitrogen. Reactive oxygen and nitrogen species (RONS) are produced by all aerobic cells and play an important role in aging as well as age-related diseases. Lipid peroxidation is a process of oxidative degradation of lipids that process by which free radicals bind to lipid electrons in the cell membrane resulting in direct cell damage. Lipid peroxidation can cause cellular damage in several ways such as impairing the integrity of the plasma membrane and subcellular organelles by peroxidation, "chain reaction" of ROS production, and activation of phospholipase A2 (PLA2) caused by lipid peroxidation. Fatty acids and other PLA2 metabolites (such as lysophospholipids) are known to damage cell membranes. In the development of kidney damage, the process of lipid peroxidation plays an important role. This is presumably due to the large number of long-chain polyunsaturated fatty acids (PUFAs) in the lipid composition of the kidneys and there are substantial evidence to suggest that ROS is involved in the ischemic, toxic, and immunologically mediated pathogenesis of renal injury, but the cellular mechanisms that result in cell injury and death are still being studied.

Keywords: oxidative stress, aging, kidney

1. Introduction

Kidney is one of the tissues affected by age. Aging involves cellular and structural changes inside the kidney and notably implicates with comorbidity, related to cardiovascular disease aging [1, 2]. The Centers for Disease Control and Prevention estimated that 72 million people in the United States (20% of population) will be 65 years old or more at 2030. Eurostat estimated, around 28% Europeans will be 65 years old at 2060. The increasing number of older adults will be predicted to increase aging related kidney disorders as well [3].

Aging kidney causes the elderly susceptible to clinical deterioration from ordinary stimulation that younger individual can compensate, including acute renal injury, volume depletion or overload, sodium and potassium level disorders, and toxic reaction against kidney excreted drugs [4]. Deficiency of kidney function is associated with death in all populations [2]. Several studies have shown that the decrease of function related to be structural (glomerulosclerosis, tubular atrophy and interstitial fibrosis) and functional (reduced glomerular filtration rate, proteinuria, decreased ability to concentrate or dilute urine, electrolyte imbalance and ion transport disorders, changes in hormonal function, reduced drug excretion) [1, 5].

As one of the organs with the fastest aging rate, kidney shows several age-related decline in both structural and functional. The renal parenchyma decreases about 1%, and creatinine clearance or GFR decrease is about 1.0 mL/minute per 1.73 m2 per year in elderly subjects [6]. In normal aging kidneys, 30% of the glomerulus are damaged and represent diffuse glomerular sclerosis by age 75. Meanwhile the remaining glomerulus denote impaired filtration ability [2]. This could explain why the prevalence of chronic kidney disease (CKD) and end-stage renal disease are very common in the elderly [6]. Chronic kidney disease is a major growing health and economic burden. About 8–13% of the world's population suffers from CKD [7].

2. Age-related kidney changes

Age-related loss of kidney function has been recognized for decades. The cross-sectional population-based study by The National Health and Nutrition Examination Survey supports the theory of age-related decline in kidney function, although some other subjects did not have an absolute decline in kidney function. Although the rate of this decline is low, the process may have negative effects on many organ systems and thus reduce overall health and physical function in the elderly individual [6]. Epidemiological, clinical, and molecular evidence suggest that aging is a major contributor to the increased incidence of acute kidney injury and CKD. Renal function recovery after an episode of acute kidney injury is significantly worse in elderly patients. Reduction of regenerative potential, which is a feature of the aging process, may be caused by aging cells [8].

With increasing age, many individuals show progressive reductions in glomerular filtration rate (GFR), renal blood flow (RBF), and loss of nephron function with wide variability between individuals [9]. The aging kidney undergoes complex changes that affect the pathology of the kidney. The underlying molecular mechanisms could be the target of future therapeutic strategies [8].

The accumulation of old cells explains the ineffectiveness of cell repair and the loss of functional ability wherein studies have shown that the removal of old cells results in delayed kidney aging. Other potential mechanisms are autophagic changes responding to renal stress and the inflammatory response [8].

3. Kidney aging and oxidative stress

Aging is a natural biological process, progressive and inevitable characterized by a gradual decline in cellular function as well as progressive structural change of organ systems. This reduction means a reduced capacity to maintain homeostatic control over essential functions and ultimately results in the death of the organism [5, 10]. These anatomical and physiological changes are called senescence, a term that describes changes more related to age than changes caused by disease [11].

The aging mechanism does not stand alone but involves the interaction of several factors. These factors were put forward by Fontana and Klein: (1) Oxidative stress will damage protein and DNA, this condition is also followed by a decrease in the ability to repair DNA, and an imbalance in the mitochondrial and nucleus genome; (2) The existence of chronic inflammation which increases with age; (3) Changes in fatty acid metabolism, which acid is associated with insulin resistence. So that there is excess free fatty acids in plasma; (4) There is an interruption of normal cell physiology due to excessive excess of metabolic and product such as Advanced glycolysis end products (AGE), amyloid, and other proteins; (5) the sympathetic nervous system and the activation of angiotensin system and changes in the neuro-endocrine system; (6) post-mitotic cell loss, resulting in a decrease in the number of neurons and muscle cells and damage to cell structure and function in all tissues and organs [12].

The kidneys have higher activity (highly energetic) than other organs, thus they produce a lot of free radicals as intermediate substances (as free radical stores) which cause oxidative stress which results in impaired signal delivery, increased apoptosis, decreased ability to regenerate cells, and fibrosis in the kidneys. This characteristics make kidney one of the organs that ages more rapidly than the other organs [13].

In aging kidney, there are interactions of genetic factors, environmental changes, and cellular dysfunction that lead to the typical structural and functional changes. The exact biological and cellular mechanisms responsible for aging are not yet known. One of the most popular theories of aging is the theory of free radicals or oxidative stress. This theory is based on the fact that cells are under chronic oxidative stress due to an imbalance between pro oxidants and antioxidants [14, 15].

This theory states that aging is caused by accumulated oxidative damage. As organisms getting older, they produce more free radicals, some of which are not completely neutralized by the endogenous antioxidant defense mechanisms. These free radicals then react with biomolecules and cause the accumulation of toxic oxidative products that lead to oxidative stress and accumulation of toxic oxidative products [10].

The kidneys depend on aerobic metabolism and oxidative phosphorylation for the production of the energy required for tubular reabsorption [16]. Due to their high metabolism, the kidneys are very susceptible to oxidative damage, and several trials have shown that oxidative stress can cause or accelerate the progression of kidney disease and its complications [17]. Most of the age-dependent renal changes such as excessive fibrosis, general lack of regenerative ability, and increased apoptosis in cells that determine healthy kidney function are often associated with excess OS [18].

The mechanism of free radicals damaging cells can be in several ways: (1) free radicals will damage the double layer lipid membrane causing disruption of the fluidity and permeability properties of the membrane. (2) Free radicals enhance protein cross-linking processes, especially those mediated by sulfhydryl groups, resulting in degradation and loss of cell activity. (3) DNA damage induced by free radicals causes the breakdown of DNA strands. (4) Impaired function of cellular receptors, neurotransmitters and hormonal responses due to oxidative damage to carbohydrates [17].

Research shows that in the aging process of organisms, excessive oxidative stress can activate many pro inflammatory pathways, including the NF-κB signaling pathway. Oxidative stress also induces ongoing regulation (chronic) of pro inflammatory mediators (such as TNF-α, IL1-β, IL-6, COX-2, iNOS) which lead to tissue and organ aging [19]. Besides, oxidative stress is also involved in various disease processes and degenerative syndromes. In mild oxidative stress due to normal metabolism results; The resulting biomolecular damage cannot be completely repaired or eliminated by cell degradation systems, such as lysosomes, proteasomes, and cytosol and mitochondrial proteases [12].

In oxidative stress, oxygen-free radicals are formed excessively as well as H_2O_2, so that the body's protection system such as catalase and glutathione peroxidase enzymes can no longer neutralize all oxygen free radicals that are formed. Furthermore, if H_2O_2 reacts with Fe^{2+} and Cu^{2+}, hydroxyl free radicals are formed through the Fenton and Haber-Weiss reaction. The hydroxyl radical is a very reactive species.

Reactive oxygen species and reactive nitrogen species can attack all types of biomolecules, especially unsaturated fatty acids which are important components of phospholipids making up cell membranes. This accumulation of oxidative damage is considered to be an important mechanism underlying aging and an increase in age-related pathology, as well as a progressive decrease in the functional efficiency of various cellular processes [14, 15]. Among the ROS, the most widely known, are superoxide ion ($O2^{\bullet -}$), hydrogen peroxide (H2O2), and peroxyl radicals ($OH^{\bullet}$), while RNS is nitric oxide (NO) and peroxy nitrite ($ONOO^-$) (**Figure 1**) [18]. At the molecular level, increased oxidative damage and its products have been reported. This contributes to a chronic inflammatory response with the accumulation of macrophages and lymphocytes in the interstitium [1].

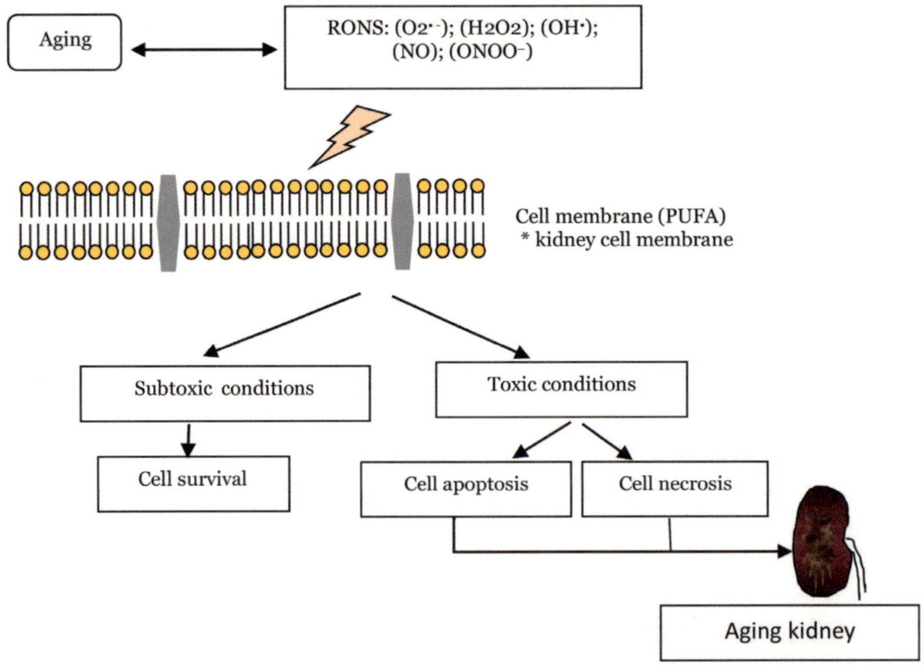

Figure 1.
Schematic mechanism of cell damage according to the theory of free radical aging. In this case the kidney.

In certain conditions that are triggered by oxidative stress, there is an insufficiency of endogenous antioxidants, both enzymatic and non enzymatic. Enzymatic antioxidants include glutathione peroxidase, catalase, and superoxide dismutase. Non enzymatic antioxidants include vitamin E, vitamin C, thiol antioxidants (glutathione, thioredoxin, and lipoic acid), melatonin, carotenoids, natural flavonoids, and others [7, 20]. Studies on old mice kidneys support the theory that there is a decreased antioxidant capacity and decreased levels of Cu/Zn-SOD, catalase, and GSH reductase.

Several processes can suppress oxidative stress: (1) reduce radiation and pollutant that affect the environment; (2) enhance antioxidant both endogen and exogen to neutralize ROS before it oxidizes cells; (3) suppress oxidative stress formation through stabilization of production and efficiency of energy in mitochondria [12].

4. The source of ROS formation

Reactive oxygen species are oxygen-derived oxidizing compounds that are highly reactive, consisting of free radicals and non-radicals. Apart from oxygen derivates, free radicals also come from nitrogen derivatives. Free radicals are atoms or molecules with unpaired electron, which are very unstable and very reactive. Therefore, they try to pair up with the free electrons around them, so that they will form other free radicals or paired electrons, where their radical characteristics may be lost. If a new free radical is formed, it is also unstable and react with other molecules to produce other free radicals or non-radical molecules due to the electron pair of the new formed molecule. Thus, a free radical chain reaction will form excessive free radicals.

Free radicals include superoxide anion (O_2), Hydroxyl radicals (OH^-), hydroperoxyl (HO_2) Peroxyl radicals (RO_2), alkoxyl (RO), carbonate ($CO3^-$), nitric oxide (NO), nitrogen dioxide (NO_2), peroxy nitrite ($ONOO^-$), nitroxyl ion, purine radical. Meanwhile, non-radicals are hydrogen peroxide (H_2O_2), organic peroxides (ROOH), ozone, oxygen siglet, and lipidperosides.

Free radicals are formed in cells through oxidation and reduction of one electron. If the formed free radicals are the oxygen's derivative, they are called reactive oxygen species (ROS). Reactive nitrogen species (RNS) are those derived from nitrogen. All of aerobic cells will produce reactive oxygen and nitrogen species (RONS) which have critical role in cell's aging and age-related diseases. The formation of RONS is not only limited to its deleterious effects but also plays a role in energy extraction from organic molecules, immune defense, and signaling processes [18, 20].

Sources of RONS can be endogenous and exogenous. Endogenous sources of RONS are generated by several major enzymatic processes such as cellular respiration (by the mitochondrial electron transport chain) or the activity of the nicotinamide adenine dinucleotide phosphate (NADPH) oxidase enzyme complex [7, 20]. Approximately 1–4% of the oxygen metabolized by mitochondria is converted to superoxide ions which in turn produce free radicals capable of damaging structural proteins and DNA [12]. Exogenous generated RONS are induced by external factors, such as chemical pollutants, environmental pollutants, such as cigarette smoke, some drugs, exposure to UV rays, alcohol, ionizing radiation, pesticides, ozone [12, 20]. X-rays and ultraviolet rays can break down water to form radicals (OH^-). Metal ions such as Fe^{2+}, Ca^{2+}, Cu^+, can also react with oxygen or hydrogen peroxide to produce OH radicals.

Among the ROS, the most widely known are superoxide ions ($O_2^{\bullet-}$), hydrogen peroxide (H_2O_2), and peroxyl radicals ($OH^{\bullet}$), while RNS is nitric oxide (NO) and peroxy nitrite ($ONOO^-$) [18]. In general, free radicals are needed for the continuity of physiological processes, especially for electron transport, cell development, as well as helping leukocytes in destroying germs in the immune system. When free radical products are balanced with the antioxidant capacity, it directs cells into growth, signaling and survival.

If these free radical products exceed the antioxidant capacity, cells will lead to oxidative stress, apoptosis and necrosis [18]. Mitochondria play an important role in the aging process because these organelles are the main source of free radical production as a side effect of energy formation. Mitochondrial dysfunction has long been considered a major contributor to aging and age-related diseases. In young mitochondria, this condition can be overcome by the presence of antioxidants in the body. However, the old mitochondria do not produce enough of the antioxidants needed to neutralize the free radicals that are formed, so there is an imbalance between the production of free radicals and antioxidants. Apart from that, mitochondria also play an important role in amino acid and lipid metabolism, calcium homeostasis, apoptosis regulation, cell cycle regulation, and thermogenesis [15, 21]. Renal fibrosis results from loss of ATP production due to mitochondrial dysfunction, which is associated with increased free radical formation and oxidative stress. Thus, oxidative stress, which involves various reactive oxygen and nitrogen species, has an important role in the pathophysiology of CKD [16].

Dysfunction of the mitochondrial electron-transport system leads to increased production of ROS, which results in mtDNA damage followed by mutations that lead to impaired mitochondrial protein function and a further increase in RONS production [21]. Electron leakage causes the reduction of one oxygen electron to form superoxide anions which are the precursors of most ROS and mediators in oxidative chain reactions. The superoxide anion then undergoes dismutation which is catalyzed by superoxide dismutase to produce durable hydrogen peroxide and a permanent membrane. This molecule can be reduced entirely or partially to water or hydroxyl radicals. Under normal conditions, ROS is maintained at a physiological level by several systems of endogenous antioxidant enzymes, such as SOD, catalase, glutathione peroxidase and glutathione reductase [17].

5. Lipid peroxidation

Oxidative stress causes tissue damage including the kidneys by a variety of different mechanisms, specifically damaging cell membranes, DNA damage, and protein modification. The term lipid peroxidation is a process of oxidative degradation of lipids. Lipid peroxidation is a process by which free radicals bind to lipid electrons in the cell membrane resulting in direct cell damage. The kidneys are highly susceptible to damage caused by ROS, possibly due to the large number of long-chain polyunsaturated fatty acids (PUFAs) in the lipid composition of the kidneys [18]. Allylic hydrogen in PUFA is very sensitive to free radical attack [14].

In response to membrane lipid peroxidation, depending on the specific cellular metabolic state and repair capacity, cells may increase cell survival or cause cell death. In conditions where the level of lipid peroxidation is physiological or low (called the subtoxic state), cells stimulate their maintenance and survival through constitutive antioxidant defense systems or activation of signaling pathways that regulates antioxidant proteins resulting in adaptive stress responses. Conversely, under moderate or high levels of lipid peroxidation (toxic conditions), the rate of oxidative damage exceeds the repair capacity. This causes programmed cell death, apoptosis or necrosis; both processes eventually cause molecular cell damage which can facilitate the development of various pathological conditions, and also accelerate aging [22].

Lipid peroxidation plays an important role physically because it decreases membrane fluidity, thereby facilitating the exchange of phospholipids between two monolayers, and increases the leakage of the bilayer membrane to substances that do not normally cross the membrane other than through certain channels [14]. Lipid peroxidation is followed by oxygen release, which is reduced into water via the mitochondrial respiratory chain. At the same time, lipids can be oxidized by efficient ROS initiators, particularly hydroxyl radicals and dihydroxyl radicals ($HO2^•$), forming water and lipid radicals. This process leads to initiation of lipid peroxidation reactions, which are constantly occurring in the cell [23].

Lipid peroxidation is an autocatalytic radical process which consists of three stages; the first is initiation, followed by propagation, and the last is the cessation of peroxidation which is the result of lipid radical interactions and/or the formation of non-radical species due to the action of peroxyl radicals. Lipid peroxidation is primarily initiated by hydroxyl radicals, which are generated through reactions catalyzed by transition metals, such as the Fenton reaction [24].

At the initiation step of lipid peroxidation, prooxidants such as hydroxyl radicals (OH^-) abstract allylic hydrogen to form carbon-centered lipid radicals ($L^·$). In the propagation phase, the lipid radical ($L^·$) rapidly reacts with oxygen to form peroxy lipid radical ($LOO^·$). Lipid peroxyl radicals are unstable molecules and can combine with other fatty acids (LH) nearby to form different lipid hydroperoxides (LOOH) and lipid radicals. Lipid peroxyl radicals can also react with themselves. The lipid hydroperoxides can then be broken down into alhoxyl lipid radicals ($LO^•$) and hydroxyl radicals ($OH^•$). The lipid radicals formed in the previous stage can react with oxygen to produce other lipid peroxyl radicals, and so on. So, this process is called the "lipid peroxidation chain reaction" (**Figure 2**) [22–24].

Once formed, LOOH can undergo reductive degradation which reduces or increases the cytotoxic potential, depending on the underlying conditions. In addition, LOOH or other intermediate peroxidation products can trigger signal transduction pathways requiring greater cytoprotection shelter (eg. upregulation of detoxification enzymes) or planned termination (apoptotic death) [25]. Lipid hydroperoxides are fairly stable compounds, but their decomposition can be catalyzed by transition metals and metal complexes, giving rise to new radicals capable of stimulating further lipid peroxidation or the formation of oxidation end products with various toxicities, such as malondialdehyde (MDA), hydroxynonenal

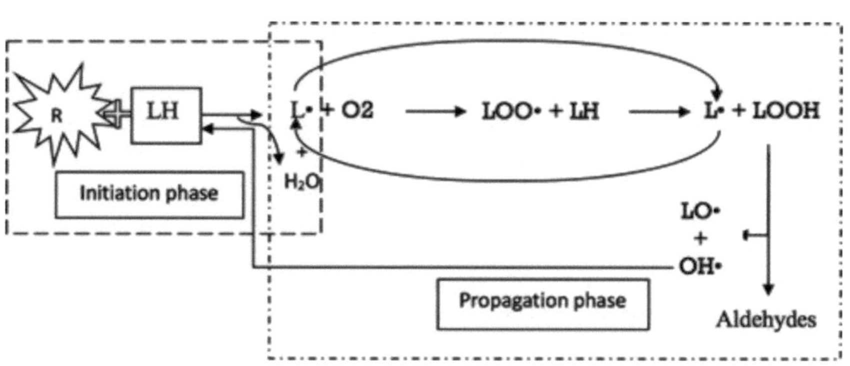

Figure 2.
Schematic reaction of the lipid peroxidation chain. LH^- lipid molecules, $L^{•-}$ lipid radicals, $LOO^{•-}$ lipid peroxyl radical, $LOOH^-$ lipid hydroperoxide, $LO^{•-}$ lipid alhoxyl radical, $OH^{•-}$ hydroxyl radical, $HO2^{•-}$ Perhydroxyl radical. The stopping phase occurs until the substrate runs out or the process is stopped by antioxidants.

and hexanal. This reaction will continue until the substrate runs out, or the process is interrupted by antioxidants [24].

In the stopping phase, an antioxidant such as vitamin E donates a hydrogen atom to the LOO˙ species and forms the associated vitamin E radical which reacts with other non-radical LOO˙ forming products. After lipid peroxidation has started, a multiplication chain reaction will continue until a stopping product is produced [22, 23].

Lipid peroxidation produces a wide variety of oxidation products. The primary product of lipid peroxidation is lipid hydroperoxide (LOOH). Secondary products of lipid peroxidation include malondialdehyde (MDA), propanal, hexanal and 4-hydroxynonenal (4-HNE). Among these secondary products, MDA appears to be the most mutagenic lipid peroxidation product, whereas 4-HNE is the most toxic [22]. Peroxidation products are involved in many cellular processes including cell metabolism, signaling, and cell survival. Lipid molecules, particularly PUFAs and cholesterol, undergo variable oxidation rate initiated by RONS [26].

6. Lipid peroxidation and its effects on the kidneys

Aging is associated with increased oxidative stress. Most of the changes in the kidneys are age dependent, such as excessive fibrosis, lack of regenerative ability in general, and increased apoptosis. At the molecular level there is an increase in mutations in both nuclear and mitochondrial DNA (mtDNA), increased lipofuscin, AGEs, oxidative stress and apoptosis. Proximal tubular cells contain a large number of mitochondria and are most dependent on oxidative phosphorylation and are most susceptible to apoptosis, oxidant-induced mutations. Recent studies have shown that the anti-aging gene, klotho, is an important factor in kidney aging and kidney damage due to oxidative stress [18].

Lipid peroxidation can cause cellular damage in several ways. First, the integrity of the plasma membrane and subcellular organelles are impaired by peroxidation. Second, the interaction of ROS with PUFAs leads to the formation of additional radicals (hydroperoxide and hydroperoxide metabolites) which result in a "chain reaction" of ROS production. This process increases the production of ROS which can cause cell damage by interacting with cellular proteins and DNA. Finally, lipid peroxidation causes activation of phospholipase A2 (PLA2). Fatty acids and other PLA2 metabolites (such as lysophospholipids) are known to damage cell membranes. So PLA2-induced free fatty acid release can cause additive injury to the cell membrane [27].

In the development of kidney damage, the process of lipid peroxidation plays an important role [28]. The kidneys are organs that are highly susceptible to damage caused by ROS. This is presumably due to the large number of long-chain polyunsaturated fatty acids (PUFAs) in the lipid composition of the kidneys [18]. There are substantial evidence to suggest that ROS is involved in the ischemic, toxic, and immunologically mediated pathogenesis of renal injury [29]. Experimentally it has been shown that ROS plays a role in the pathogenesis of kidney disease, but the cellular mechanisms that result in cell injury and death are still being studied [27].

In general, human and animal studies suggest that lipid oxidation plays an important role in predicting the development of cardiovascular and renal disease and the response to therapy. In kidney disease, the use of biomarkers such as malondialdehyde (MDA), isoprostanes (IsoPs) or isolevuglandins (IsoLGs) has been reported. Malondialdehyde can interact with proteins and can potentially cause atherogenicity [7].

ROS causes lipid peroxidation in cell membranes and organelles, thereby impairing structural integrity and capacity for cell transport and energy production, particularly in the proximal tubular segment. In addition, lipid peroxidation co-mediates decreased glomerular blood flow and glomerular filtration through release of vasoconstrictive bioactive lipids (prostaglandins, thromboxanes, and platelet activating factors) and, possibly, nitric oxide relaxation inactivation [29].

Li et al's study of 2,169 adult patients stated that oxidative stress can be demonstrated by plasma MDA levels associated with the prevalence of mild acute renal insufficiency and/or CKD. Although MDA increases the load on the kidneys and/or causes oxidative stress cycle in the body, high levels of MDA in plasma may be associated with age-related decline in kidney function as well [6].

Gomes et al. examined renal oxidative stress status in old Winstar Kyoto (WKY) mice. This research was conducted by measuring H2O2 levels in the kidneys. In this study, it was found that the renal and medullary cortical H2O2 production increased sharply in old WKY. This suggests that the dramatic increase in the rate of H2O2 production in the old WKY kidney is indicative of a significant increase in oxidative stress in this tissue. Moreover, increased renal ROS production and lipid-related oxidative damage may play a role in the pathogenesis of kidney disease. Overall, these data suggest that elderly WKY exhibits the first signs of renal oxidative damage [5].

Lim et al's study on old C57/BL6 mice showed that aging kidneys exhibited increased levels of reactive oxygen species and thiobarbiturate acid reactive substances, which are associated with oxidative lipid damage. In addition, other markers of oxidative stress and lipid peroxidation, such as isoprostane, AGE, and elevated heme oxygenase, were also found in old mice. This research also shows that Sirtuin 1 and Klotho also decrease with aging [30].

Antioxidant enzymes are able to remove reactive oxygen species and lipid peroxidation products. In aging, physiological functions change due to a decrease in endogenous antioxidants, such as SOD, CAT, and GSH-Px. In contrast MDA, a good indicator of lipid peroxidation showed an increase. Chen et al. in their study of old mice showed that the activity of SOD, CAT and GSH-Px in the liver and kidneys decreased compared to the group of young mice. In addition, MDA levels in the liver and kidneys of old rats were increased compared to the group of young mice [31].

Besides the endogenous antioxidants above, there is also a role for enzymes involved in oxidative stress response, aging and various metabolic regulation in the body, namely Sirtuin1. This enzyme is known as the master regulator. Starting with the discovery of the silent information regulator 2 (Sir2) gene in yeast in 1986. Where the long-lived yeast has an overexpression of this gene, while in the shorter-lived yeast it is found to have low expression. Since then Sir2 is believed to be a gene that plays a role in longer survival (longevity) [32, 33]. Homologous Sir2 gene in mammals, is sirtuin (SIRT).

Sirtuin is a class III histone deacetylase protein group that has deacetylation activity against histones and non-histones that do not contain nicotinamide adenine dinucleotide (NAD) deacetylase and/or adenosine diphospate (ADP) ribosyltransferase. This enzyme works with the help of the coenzyme NAD to carry out its function [34]. Sirtuin 1 (SIRT1) is considered to be most homologous to yeast sir2. In the study of young mice, Sirtuin1 is expressed by various cells in organs such as kidneys, liver, lymph, skin, but mostly in the kidneys, especially the medulla [35–38].

Sirtuin 1 not only has deacetylation activity against histones but also on many transcription factors and cofactors, such as p53, FOXO, peroxisome proliferator activated receptor γ (PPAR-γ), co-activator-1α and NF-κB which play a role in crucial cellular activity including the response to stress, metabolism and longevity (cell senescence) [39, 40].

As a function of redox regulator, sirtuin detects surrounding imbalance through NAD levels, sirtuin 1 will deacetylase the substrate so that it will activate antioxidant genes such as SOD2 (superoxide dismutase 2), GPX1 (glutathione peroxidase 1) which can anticipate free radicals. Under conditions of oxidative stress, sirtuin can produce O-acetyl-ADP ribose (OAADPR) which can turn into ADP ribose, both of which have a protective effect against oxidative stress [41].

The role of sirtuin through the Forkhead box (FOXO) is to regulate apoptosis, lipid metabolism, cell proliferation, inflammation, autophagy and stress resistance. Where FOXO3 acetylation expresses apotosis-related genes, such as: Bim, TRAIL and FasL [33]. Nuclear factor-κB (NF-κB) is a widespread transcription factor affecting inflammation, apoptosis, adhesion and the cell cycle through regulation of target genes. Sirtuin can inhibit the activity of NF-κB so that it can play a role in glucose control, AGEs, cytokines, growth factors, dan toll-like receptors [42].

Sirtuin 1 regulates lipid homeostasis by regulating the sterol regulatory element binding protein (SREBP), liver X receptor (LXR) and farnesoid x receptor (FXR). Sirtuin 1 directly deacetylates SREBP, inhibits SREBP target gene expression and reduces lipid and cholesterol levels [33, 34]. Tumor suppressor p53 is a transcription factor associated with oxidative stress responses. With the ability of SIRT 1 to deacetylate P53 it will reduce transcription activity.

Increased oxidative stress in the elderly will decrease SIRT1 levels, resulting in increased inflammation, increased apoptosis, decreased autophagy, decreased levels of endothelial nitric oxide synthase (eNOS), increased AT1R expression, reduced ability in redox reactions and lipid metabolism. The reduced expression of SIRT1 due to aging will make the kidneys prone to progressive structural and functional disorders (**Figure 3**) [12, 33, 43].

Kume et al. (2010) found the association between mitochondrial damage of proximal tubular cells in aging kidneys with reduced expression of sirtuin. In proximal tubular cells of old mice, autophagy responses to kidney hypoxia were decreased and caused dysfunction and fibrosis. Increasing sirtuin expression by caloric restriction helps improvement of aging kidneys [44]. The similar result was stated by He et al. (2010), who found sirtuin expression protected kidneys in oxidized state and provided anti apoptotic and anti fibrotic effects to kidneys [45].

Kidney function is determined by glomerular filtration rate. This filtration barrier consists of endothelial cells, glomerular basal membrane and podocytes. Podocyte is critical in maintaining glomerular filtration. Podocyte damage caused

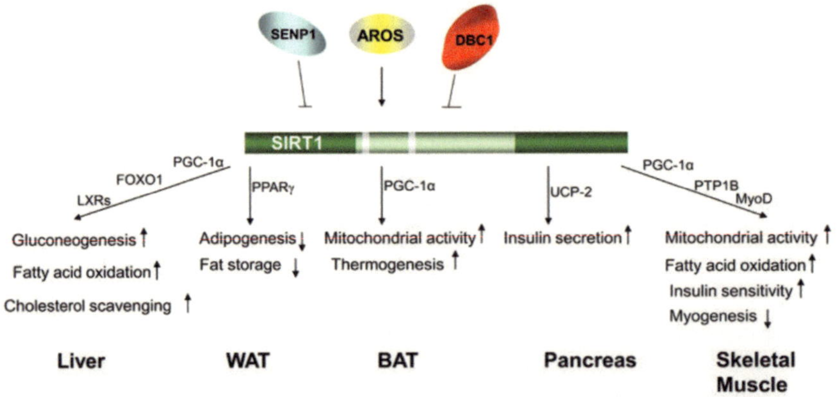

Figure 3.
The role of SIRT1 in metabolic Homeostais [42].

by oxidative stress leads to reduction of glomerular filtration rate. Sirtuin preserves homeostasis and protects podocytes from oxidation.

Transforming growth factor (TGF)-β is a key cytokine that regulates apoptosis, cell cycle, differentiation and accumulation of extracellular matrix. Association between TGF-β/Smad and kidney fibrosis occurrence has been proven in many studies. Sirtuin, which suppress expression of TGF-β, is decreased in aging kidneys. This results in increased expression of TGF-β thus also increases kidney fibrosis.

Author details

Harnavi Harun
Nephrology Division, Internal Medicine Department, Medical Faculty of Andalas University, West Sumatera, Indonesia

*Address all correspondence to: harnavi@med.unand.ac.id; hharnavi19@gmail.com

References

[1] Braun F, Rinschen MM, Bartels V, Frommolt P, Habermann B, Hoeijmakers JHJ, et al. Altered lipid metabolism in the aging kidney identified by three layered omic analysis. Aging. 2016; Vol. 8 No.3: 441-454.

[2] Kanasaki K, Kitada M, Koya D. Pathopysiology of the aging kidney and therapeutic interventions. Hypertension Research. 2012; 35:1121-1128.

[3] O'Sullivan ED, Hughes J, Ferenbach DA. Renal Aging: Causes and Consequences. J Am Soc Nephrol. 2017; 28:407-420.

[4] Musso CG, Oreopoulus DG. Aging and Physiological Changes of the Kidneys Including Changes in Glomerular Filtration Rate. Nephron Physiol 2011;119 (suppl):p1-p5.

[5] Gomes P, Simão S, Silva E, Pinto V, amaral JS, Afonso J, et al. Aging increases oxidative stress and renal expression of oxidant and antioxidant enzymes that are associated with an increased trend in systolic blood pressure. . s.l. : Oxidative Medicine and Cellular Longevity. 2009; 2(3): 138-145.

[6] Li G, Chen Y, Hu H, Liu L, Hu X, Wang J, et al. Association Between Age-Related Decline of Kidney Function and Plasma Malondialdehyde. Rejuvenation Research. 2012; 15, 257-264.

[7] Krata N, Zagożdżon R, Foroncewicz B, Mucha K. Oxidative Stress in Kidney Diseases: The Cause or the Consequence? Archivum Immunologiae et Therapiae Experimentalis. 2018; 66.211-220.

[8] Schmitt R, Melk A. Molecular mechanisms of renal aging. Kidney International. 2017; 92:569-579.

[9] Weinstein JR, Anderson S. The Aging Kidney: Physiological Changes. Adv Kidney Dis. 2010; 17(4): 302-307.

[10] Pratic D. Lipid Peroxidation and the Aging Process. Sci. Aging Knowl. Environ. 2002; 50: p. re5.

[11] Denic A, Glassock RJ, Rule AD. Structural and functional changes with the aging kidney. Adv Chronic Kidney Dis. 2016; 23 (1): 19-28. Doi:10.1053/j.ackd.2015.08.004.

[12] Poljsak B, Milisav I, Aging Oxidative Stress and Antioxidants, in: Oxidative Stress and Chronic Degenerative Dseases – A Role for Antioxidants. IntechOpen. 2013; 332-353.

[13] Perico N, Remuzzi G, Benigni A. Aging and The Kidney.Current Opinion in Nephrology and Hypertension. 2011; 317-321.

[14] Ward WF, Qi W, Remmen HV, Zackert WE, Roberts II LJ, Richardson A. Effects of age and caloric restriction on lipid peroxidation: measurement of oxidative stress by F2-Isoprostane Levels. Journal of Gerontology: Biological Sciences. 2005; Vol. 60A, No. 7, 847-851.

[15] Harun H, Yanwirasti Y, Purwanto B, Rahayuningsih EP. The Effect of Giving Dadih on Malondialdehyde Levels and Renal Interstitial Fibrosis at Aging Kidney. Open Access Macedonian Journal of Medical Sciences. 2020 May 15; 8 (A): 293-296.

[16] Lv W, Booz GW, Fan F, Wang Y, Roman RJ. Oxidative Stressand Renal Fibrosis: Recent Insights for the Development of Novel Therapeutic Strategies. Front. Physial. 2018. 9:105. doi: 10.3389/fphys.2018.00105.

[17] Pellefrino D, La Russa D, Marrone A. Oxidative Imbalance and Kidney Damage: New Study Perspectives from Animal Models to Hospitalized Patients. Antioxidants. 2019; 8, 594.

[18] Ozbek E. Induction of Oxidative Stress in Kidney. International Journal of Nephrology. 2012;9 pag.

[19] Xi Y, Wang M, Zhang W, Bai M, Du Y, Zhang Z, S. Shin, *et al.* Neuronal Damage, Central Cholinergic Dysfuntion and Oxidative Damage correlate with Cognitive Deficits in Rats with Chronic Cerebral Hypoperfusion. Neurobiol Learn Mem. 2014. 109: 7-19

[20] Liguori I, Russo G, Curcio F, Bulli G, Aran L, Della-Morte D, et al. Oxidative stress, aging, and diseases. Clinical Interventions in Aging. 2018;13: 757-772.

[21] Cedikova M, pitule P, Kripnerova M, Markova M, Kuncova J. Multiple Roles of Mitochondria in aging Processes. Physial. Res. 2016;65(Suppl. 5): S519-S531.

[22] AyalaA, MuňozMF, ArgüellesS. Lipid Peroxidation: Production, Metabolism, and Signaling Mechanisms of Malondialdehyde and 4-Hydroxy-2Nonenal. Oxidative Medicine and Cellular Longevity. 2014: 31 pages.

[23] *Kudryavtseva AV, Krasnov GS, Dmitriev AA, Alekseev BY, Kardymon OL, Sadritdinova AF, et al. Mitochondrial dysfunction and oxidative stress in aging and cancer. Oncotarget. 2016; 7(29):44879-94.*

[24] *Muňiz P, Coma MJ, Terán J. Oxidative Stress and Vascular Damage In Hypoxia Processes. Malondialdehyde (MDA) as Biomarker for Oxidative Damage. Rev Electron Biomed / Electron J Biomed. 2014;2:50-53.*

[25] *Girotti AW. Lipid Hydroperoxide generation, turnover, and effector action in biological systems. Journal of Lipid Research. 1998; 39: 1529-42.*

[26] *Ademowo OS, Dias HKI, Burton DGA, Griffit HR. Lipid (per) oxidation in mitochondria: an emerging targetin the ageing process? Biogerontology. 2017; 18:859-879.*

[27] *Sheridan AM, Fitzpatrick S, Wang C, Wheeler DC, Lieberthal W. Lipid Peroxidation contributes to hydrogrn peroxide inducedcytotoxicity in renal epithelial cells. Kidney International. 1996; 49:88-93.*

[28] *Ossani G, Dalghi M, Repetto M. Oxidative damage lipid peroxidation the kidney of choline-deficient rats. Frontiers in Bioscience. 2007; 12, 1174-1183.*

[29] *Baud L, Ardaillou R. Involvement of reactiveoxygen species in kidney damage. British Medical Bulletin. 1993; 49, (3): 621-629C.*

[30] *Lim JH, Kim EN, Kim MY, Chung S, Shin SJ, Kim HW, et al. Age-Associated Molecular Changes in the Kidney in Aged Mice. Oxidative Medicine and Cellular Longevity. 2012; 10 pages.*

[31] *Chen X, Zhang Y, Yang l, Zu Y, Lu Q. Effects of rosmarinic acid on liver and kidney antioxidant enzymes, lipid peroxidation and tissue ultrastrructure in aging mice. Food Funct., 2015; 5 pages.*

[32] Wakino S, Hasegawa K, Itoh H. Sirtuins and metabolic kidney disease. International Society of Nephrology. 2015; 88(4): 691-8.

[33] Kong L, Wu H, Zhou W, Luo M, Tan Y, Miao L, et al. Sirtuin 1: A target for kidney disease. Mol Med. 2015; 21: 87-97.

[34] Kitada M, Kume S, Watanabe A, Kanasaki K, Koya D. Sirtuins and renal disease: a relationship with aging and diabetic nephropathy. Clinical Science. 2013; 124: 153-64.

[35] Guan Y. Hao CH. SIRT1 and kidney function. Kidney Diseases. 2015; 1: 258-65.

[36] Nath KA. The role of sirtuin 1 in renal rejuvenation and resistance to stress. The Journal of Clinical Investigation. 2010; 120: 1026-8.

[37] Canto C, Auwerx J, Desvergine B. Targeting sirtuin 1 to improve metabolism: All you need is NAD^+? Pharmacological Reviews. 2012; 64(1): 166-87.

[38] Kwon Y, Kim J, Lee CY, Kim H. Expression of SIRT1 and SIRT3 according to age in mice. Anat Cell Biol. 2015; 48: 54-61.

[39] Lim JH, Kim EN, Kim MY, Chung S, Shin SJ, Kim HW, et al. Age associated molecular changes in the kidney in aged mice. Oxid Med Cell Longev. 2012; 11: 1-10.

[40] Kvell K. Changes of renal function, electrolyte/water and acid/base homeostasis. In: Kvell K, Pongracz J, Szekely M, Balasko M, Petervari E, Bako G, editors. Molecular and Clinical Basic of Gerontology. Pecs: University of Pec. 2011: p63-73.

[41] Santos L, Escando C, Denicola A. Potential modulation of sirtuins by oxidative stress. Oxidative Medicine and Cellular Longevity. 2016; 1: 1-12.

[42] Yu J, Auwerx J. The role of sirtuins in the control of metabolic homeostasis. Integrative Physiology. Ann N Y acad. 2009; 1: 10-9.

[43] Hao CM, Haase VH. Sirtuins and their relevance to the kidney. J Am Soc Nephrol. 2010; 21: 1620-7.

[44] Kume A, Uzu T, Horike K, Kanasaki M, Isshiki K, Araki S, Sugimoto T, et al. Calorie restriction enhances cell adaption to hypoxia through SIRT1-dependent mitochondrial autophagy in mouse aged kidney. The Journal Clinical Investigation. 2010; 120: 1043-55.

[45] He W, Wang Y, Zhang M, You L, Davis LS, Fan H, et al. Sirt1 activation protects the mouse renal medulla from oxidative injury. The Journal of Clinical Investigation. 2010; 120: 1056-68.